T0305747

Spatial Data on Water

Series Editor
Jean-Charles Pomerol

Spatial Data on Water

Geospatial Technologies and Data Management

Hassane Jarar Oulidi

First published 2019 in Great Britain and the United States by ISTE Press Ltd and Elsevier Ltd

ISTE Press Ltd
27-37 St George's Road
London SW19 4EU
UK

www.iste.co.uk

Elsevier Ltd
The Boulevard, Langford Lane
Kidlington, Oxford, OX5 1GB
UK

www.elsevier.com

For information on all our publications visit our website at http://store.elsevier.com/

British Library Cataloguing-in-Publication Data
A CIP record for this book is available from the British Library
Library of Congress Cataloging in Publication Data
A catalog record for this book is available from the Library of Congress
ISBN 978-1-78548-312-7

Printed and bound in the UK and US

Contents

Acknowledgements

I would like to thank the staff and professors at the Hassania School of Public Works (EHTP), in particular the department of Mathematics, Informatics and Geomatics, which provided the necessary support for this project.

I thank also my colleagues for their assistance with reviews: Guy Mélard and Atika Cohen.

Thanks to my PhD students Aniss Moumen and Khazaz Lamiaa who contributed to the case studies.

Special thanks to my family, wife and kids.

Finally, I thank the ISTE team for their high-quality professional support during the preparation of the book.

List of Acronyms

AFIGEO	Association Française pour l'Information Géographique [*French Association for Geographical Information*]
AFNOR	Association Française de Normalisation en France [*French Association for Standardization in France*]
AGIRE	Programme d'Appui à la Gestion Intégrée des Ressources en Eau [*Moroccan Support Programme for the Integrated Management of Water Resources*]
AHIS	Automatic Hydrological Information System
ANSI	American National Standards Institute
API	Application Programming Interface
AWRIS	Australian Water Resources Information System
BADRE21	Base de Données des Ressources en Eau du 21ème siècle [*French Database for Water Resources in the 21st Century*]
BOM	Australian Bureau of Meteorology
CNIG	Conseil National de l'Information Géographique [*French National Council for Geographical Information*]
CORBA	Common Object Request Broker Architecture
COVADIS	Commission de Validation des Données pour l'Information Spatialisée [*French Commission for Validation of Data on Spatial Information*]
CSIRO	Commonwealth Scientific and Industrial Research Organization
CSW	Catalog Service for the Web

CUAHSI	Consortium of Universities for the Advancement of Hydrologic Science
DCOM	Distributed Component Object Model
DEM	Digital Elevation Model
DGH	Direction Générale de l'Hydraulique [*Moroccan General Directorate of Hydraulics*]
DBMS	Database Management System
DRG	Digital Raster Graphic
DRPE	Direction de la Recherche et de la Planification de l'Eau [*Moroccan Directorate for Research and Water Planning*]
EEA	European Economic Area
EMWIS	Euro-Mediterranean Information System on Know-how in the Water Sector
EPSG	European Petroleum Survey Group
ESRI	U.S. Environmental Systems Research Institute
EU	European Union
EUWI	European Union Water Initiative
FGDC	U.S. Federal Geographic Data Committee
FOSS	Free and Open-Source Software
GeoJSON	Geographic JavaScript Object Notation
GDI	Geospatial Data Infrastructure
GEOSS	Global Earth Observation System of Systems
GeoTIFF	Geographic Tag(ged) Image File Format
GIF	Graphics Interchange Format
GIN	Groundwater Information Network
GIS	Geographic Information System
GML	Geography Markup Language
GPS	Global Positioning System
GSDI	Global Spatial Data Infrastructure
GSM	Global System for Mobile Communication

GWML	GroundWater Markup Language
HBA	Hydraulic Basin Agency
HCP	Haut Commissariat au Plan [*Moroccan High Commissioner for Planning*]
HIDROWEB	Sistema d'Informaçoes Hidrologicas [*Brazilian Hydrological Information System*]
HISPAGUA	Sistema Español de Información sobre el Agua [*Spanish Hydrological Information System*]
HTML	Hypertext Markup Language
HTTP	Hypertext Transfer Protocol
IGN	Institut National de l'Information Géographique et Forestière [*French National Institute for Geographical and Forestry Information*]
India-WRIS	Water Resources Information System of India
INSPIRE	Infrastructure for Spatial Information in Europe
ISO	International Organization for Standardization
JS	JavaScript
JSON	JavaScript Object Notation
JSP	Java Server Pages
KML	Keyhole Markup Language
MRE	Ministère des Ressources en Eau [*Algerian Ministry of Water Resources*]
NDVI	Normalized Difference Vegetation Index
NetCDF	Network Common Data Form
NICT	New Information and Communication Technologies
OGC	Open Geospatial Consortium
ONEMA	Office National de l'Eau et des Milieux Aquatiques [*French National Office for Water and the Water Environment*]
PHP	Hypertext Preprocessor
PISEAU	Programmes d'Investissement du Secteur d'Eau [*French Water Sector Investment Program*]

PNG	Portable Network Graphics
RAOB	Réseau Africain des Organismes de Bassin [*African Network for Catchment Basin Organizations*]
RNDE	Réseau National des Données sur l'Eau [*French National Network for Water Data*]
SADIEau	Système Africain de Documentation et d'Information sur l'Eau [*African Water Documentation and Information System*]
SANDRE	Service d'Administration Nationale des Données et Référentiels sur l'Eau [*French National Service for Water Data and Common Repositories Management*]
SDSS	Spatial Decision Support System
SINEAU	Système d'Information National d'EAU [*Tunisian National Water Information System*]
SiSOL	Service d'Instrumentation Sol [*French Soil Instrumentation Service*]
SLD	Style Layer Descriptor
SNDE	Schéma National des Données sur l'Eau [*French National Program for Water Data*]
SOA	Service-Oriented Architecture
SOS	Sensor Observation Service
SQL	Structured Query Language
SYGREAU	Système de Gestion des Ressources en Eau [*Tunisian Water Resources Management System*]
UfM	Union for the Mediterranean
UN	United Nations
UML	Unified Modeling Language
UNEP	United Nations Environment Program
UNESCO	United Nations Educational, Scientific and Cultural Organization
URL	Uniform Resource Locator
USGS	United States Geological Survey
WaterML	Water Markup Language
WCS	Web Coverage Service

WDTF	Water Data Transfer Format
WFS	Web Feature Server
WIFI	Wireless Fidelity
WISA	Water Information System for Austria
WIS	Water Information System
WISE	Water Information System for Europe
WKB	Well-Known Binary
WKT	Well-Known Text
WMC	Web Map Context
WMO	World Meteorological Organization
WMS	Web Map Service
WPS	Web Processing Service
XML	Extensible Markup Language

Preface

Data around water, whether it is hydrologic, meteorological or hydrogeological, comprises a considerable wealth, hence the high cost of its acquisition, production and updating. However, it is not always managed in an optimum manner.

On the one hand, this is linked to the lack of knowledge as to its existence and its localization, the heterogeneity of technologies and tools used, which makes sharing, research, access, interpretation and the use of this data difficult. On the other hand, it is linked to the existence of a multitude of players that are involved in data collection and management. This leads to data redundancy with various types, formats and qualities. These factors pose problems of both access and use of data coming from multiple organizations, thereby limiting the planning and the use of this data for decision-taking, which varies according to the given time, scale, structure and participants, who can be policy-makers, planners, professionals or researchers.

In this context, geospatial technologies play a major role as regards scientific research, assisting decisions for the integrated and sustainable management of water resources, as well as for the contribution of various players in decision-making processes.

Geospatial technologies are also a powerful vector for the introduction of new information technologies, in particular favoring the sharing and the exchange of information between public services as part of the policy of modernizing the public sector. A multitude of international initiatives (such as WISE, SANDRE and eWater) were developed to this end.

This work illustrates the contribution of geospatial technologies for the better management of data around water, a preliminary stage giving concrete expression to actual integrated management of water resources. This work also touches on examples of initiatives as regards pilot development for implementing an infrastructure for spatial data on water (known in France as "IDSE").

Hassane JARAR OULIDI
October 2018

Introduction

Water as a vital resource is, from now on, essential as regards the major concerns of this century. Demographic and technological transformations associated with globalization and climate change have impacts around the questions of water linked both to issues of sustainable development and territorial governance. The Mediterranean area particularly, where the growing demand for water is associated with a reduction in this resource, requires rational and optimal management. It is thus necessary to pursue applied research activities around water management and to develop innovative practices that are based upon the integration of both the geospatial dimension and new IT and communication technologies (NICTs).

The core aspect of these technologies is their capacity to gather together within a single tool, variable and geographically localizable data. The tool is not limited to compiling and communicating information. It also enables the analysis, manipulation and management of such data, the simulation of various development scenarios and the reproduction of their results.

The wealth and availability of communication tools for geographical information corollary to the development of IT and communication networks have enabled the sustained progress within the sphere of geographical information systems (GIS). The progress that this sphere has witnessed has caused a transfer of GISs towards GIS-webs.

The publication of cartographic data on the Internet has become a necessary means of communication for the various organizations that are manipulating geoscientific information. Generally, this technology, based upon a form of architecture known by the "client/server" designation, is widely used for cartographic dissemination applications [GOG 01]. It is also implemented using free tools (open source) made available to the general public and initiated by the international scientific community. Water itself is part of geoscientific information. However, standardization remains an impediment and today mobilizes think-tanks at all levels, so as to construct a framework for interoperability, and it enables an exchange of these data [ATE 12] [OLI 05].

In relation to water data, this interoperability can only be guaranteed by the implementation of advanced information systems. We must stress that new scientific research fields are looking into the development of spatial data infrastructures dedicated to water resources, facilitating the use, sharing and the relevant exploitation of this data.

The SOA (service-oriented architecture) concept is one group of operations, which might be cited by users, enabling them access to the information and thus responding to their demands and needs. The Open Geospatial Consortium (OGC) and ISO/TC 211 both adopt this concept offering a series of specifications that use web services (WS) to the geospatial community. This thereby provides access to data and services distributed via URLs (uniform resource locators) [GIU 11]. This emphasizes the full potential of interoperability, enabling establishments operating within the spatial data production sector to publish and exchange data with other interoperable systems. The use of such OGC web services offers the possibility of interfacing such data and reusing it in a transparent way within a variety of applications. Although the current GDIs offer the possibility for data research, visualization and access, with the support of interoperable services and SOA concepts, it is now possible to construct new applications based upon distributed services. It thereby allows for the best increase in the value of hydrologic data.

On the country scale, the GDI around water provides a space for exchange and for sharing water information and data produced by the various players in the water sector. It is within a partnership framework between all interested players in this sector, whether public operators (ministries, public services, agencies, offices and others), the private sector (companies, research units and similar organizations), researchers and specialists and, more widely, the "general public". Within this framework, the GDI around water is a social project, the aim of which is to offer decision-makers, experts, scientists and the general public, not only the means to access information around water, but also to reuse and share it. Hence the significance of geospatial technologies as regards the development of technical infrastructures for water data and information sharing.

At the global level, several initiatives have been developed to enable data exchange around water, in particular the working group around water, falling within the "GEO Observation GEOSS" initiative (https://www.earth observations.org). At European level, the water information system, WISE, and the directive, INSPIRE, constitute the community framework for the sharing and the dissemination of knowledge around water. At the Mediterranean country level, the EMWIS system is a mechanism that aims to favor the exchange of water information and data collection at the catchment area level.

Scientific goal of the book

The most significant goal of this work is to make water data interoperable by using geospatial technologies enabling users to spend more time analyzing, rather than collecting data. It may also serve as a starting point for the identification and understanding of the operation of hydrologic systems. This work thus enables us to:

– make an inventory as regards the development of water information systems;

– initiate the infrastructures for spatial data;

– increase the information interoperability around water resources, produced by organizations operating within the water sector;

– disseminate highly reliable geographical background information to scientists, which may be directly usable;

– enable water specialists to share a common vision of the hydrologic environment;

– access various sources of water data directly and quickly; and

– present four case studies illustrating the contribution of GDIs as regards water data management.

Structure of the book

This book is divided into three parts: the theoretical framework, the technical framework and the presentation of four case studies.

The theoretical section sets up a general view of the concepts and constituents of a spatial data infrastructure, interoperability and the norms, as well as the standards of GI (geographic information). This introduction is followed by an overview of the required geospatial web services to put a GDI around water into practice, with a focus around OGC standards as well as aspects of technical implementation. Lastly, before concluding, there is a presentation of geospatial technologies with a focus around existing solutions within the open source market.

As for the technical framework, this book is used as a means of presenting international initiatives as regards GDI development around water resources and the various water information systems, as well as international data exchange standards. Finally, there is an architectural model which serves to put a GDI into practice, enabling relevant water data management. This analysis, both theoretical and technical, will be supported by four case studies.

Case Study 1: GDI on groundwater resources. This case study tackles the problem of developing a technical framework for the implementation of a geospatial data infrastructure (GDI) to make data relating to groundwater resources interoperable in order to better manage and preserve them. This enables the supply of reliable information to professionals and decision-makers in the field.

Case Study 2: Sensor Observation Service for sustainable water resource management. This case study sets out a framework made up of a module implementing sensor web enablement (SWE) standards, associated with a geoportal and a geocatalog. All of this architecture, having a service orientation, is based around open source solutions and conforms to OGC standards (WMS, WFS, CSW and other standards). The objective sought is to demonstrate the provision of this architecture, to produce a better form of monitoring, and monitoring key indicators within the management of water resources.

Case Study 3: geoprocessing water data. This study concerns the increase in GDI performance through the integration of the standard web processing service (WPS). Hence, this system will enable the production of online geoprocessing operations, intuitively, by simple mouse clicks, without having to install software. The ultimate aim is to enable decision-making whilst ensuring better integrated and sustainable management regarding both water and the environment.

Case study 4: designing a decision support tool. This study aims to establish a document, intended to clarify for decision-makers and town and country planners the most productive fields from a hydrological viewpoint. When faced with this document, decision-makers as well as town and country planners can define, in the short and long term, a plan of action for optimum groundwater exploitation. Finally, these cartographic documents have been integrated within a GDI around water.

Target audiences

This book, given its content, is intended for professionals working within the field of water resource management, as well as researchers, academics and students wishing to specialize in water resource management.

Theoretical Framework

In this chapter, we consider the theoretical bases that will enable readers to define the general concepts of geospatial technology and their features in the management of water resources. This theoretical framework is subdivided into three sections. The first (section 1.1) is an initiation concerning geographical information that stresses the crucial significance of the quality of geospatial data. The second section (section 1.2) examines the basic principles of a geospatial data infrastructure (GDI), discussing the significance of its interoperability as well as its various forms. We also define the norms and standards commonly used in the implementation of a GDI. The third section (section 1.3) provides an overview of present geospatial technologies in regards to the development of a GDI. A study of benchmarking is covered in detail, enabling future users to better choose between the tools available on the open-source market. Comprehending the theoretical framework is essential in assisting the reader to better understand the technical aspects and the proposed case studies to come.

1.1. Geospatial information system

The geospatial concept was, in practical terms, used towards the end of the 1980s. However, the application of spatially referenced data and its use existed well before the emergence of so-called "geospatial technologies". For a long time, people combined data with geographical positions and used it to improve their knowledge of a given field of study. Geospatial analysis is defined as the processing and interpretation of geographical information. Currently, geospatial information systems (GIS) are of huge interest to researchers and engineers operating in the field of water management. GIS as a technology is designed to store, model, manipulate, analyze and display localized data (Figure 1.1). GIS also enables us to ensure a link between geographical entities and semantic data; to combine and display vast volumes of data.

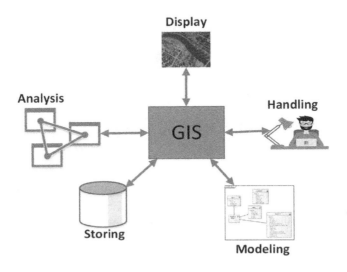

Figure 1.1. *The main components of GIS*

A large part of data that is currently available in the field, whether environmental, socio-economic or climate data, can be referenced spatially, thus offering opportunities for spatial analysis of trends and relationships. This data can be shown in two ways:

– the vector data mode represents geometric objects defined as points, lines or polygons, including three-dimensional specific data (3D). The vector formats are the type that is the most commonly used by GIS applications, as the easiest data to manipulate and the resulting files are smaller in size; and

– the raster data mode is generally used to store raster format images, such as digitized hard copy maps, aerial photographs or even satellite images such as Landsat, QuickBird, IKONOS and others.

Each mode of data representation has advantages and disadvantages for storage efficiency as well as the types of processing which can be completed. To a certain extent, the choice of mode is often completed according to the user's experience, their preferences or the nature of what these data are representing in the real world.

1.1.1. *Types of geospatial data*

The GIS software will be able to store geospatial data in several formats. A GIS data format should imperatively respond to the following elements:

– geospatial objects represent geographical and geometrical characteristics;

– metadata detail information around data, including the forecasting system used; the quality of geographical data, data acquisition techniques, the production date and the date of updating;

– semantic data supply additional information around geographical entities; and

– data on style, such as the color or the type of line to use, show geographical data on a map.

The most up-to-date vector formats comprise:

– the shapefile (*.shp): a format developed by the company ESRI (Environmental Systems Research Institute) to store and exchange GIS data. A shapefile is made up of a collection of files all having the same base file name, for example hydrography.shp, hydrography.shx (which contains the geometric index) and hydrography.dbf (containing recipient data);

– simple geographic entities: an OpenGIS OGC specification to store geographic data (for example points, lines, polygons) as well as associated attributes;

– coverage: a data format used by ESRI to store the ArcInfo information type.

Diagrammatically speaking, we can conclude that there are two data formats: a static format which stores data in formats such as those mentioned above, and a dynamic format which offers the possibility both to store data to show them within a program in the course of implementation, and to exchange geographical entities between two GIS applications.

So-called dynamic formats are:

– well-known text (WKT): a format within a text mode which serves to show a geographical entity (line, point or polygon);

– well-known binary (WKB): this takes a binary mode format rather than a textual format to represent a single geographical entity;

– GeoJSON: an open format for coding geographical data structures, based upon the data exchange format JSON; and

– Geography Markup Language (GML): an open standard XML for the exchange of GIS data.

The most currently used formats are:

– digital raster graphic (DRG): this format is used for the storage of digitized hard copy maps using an electronic scanning device (Figure 1.2);

– digital elevation model (DEM): used to store information on altitude (Figure 1.3);

– Geotiff: these data formats are generally used to store raster data. Such a format also allows you to associate georeferencing information with a tiff image.

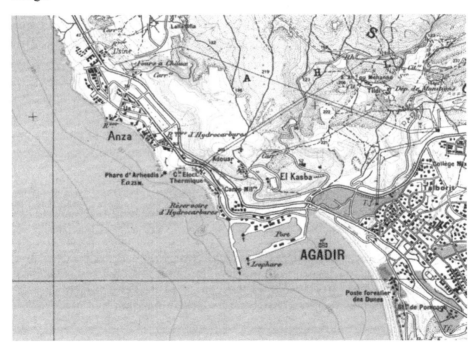

Figure 1.2. *Part of a topographical map of the town of Agadir using the DRG format. For a color version of this figure, see www.iste.co.uk/jarar/spatial.zip*

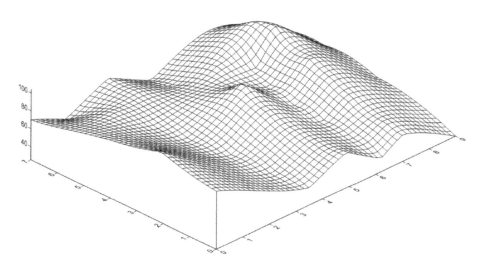

Figure 1.3. *Example of a digital elevation model (DEM)*

1.1.2. *Geographical database*

A database is partly a "data set modeling objects from a part of the real world and serving as a means of support for an IT application. To deserve the term database, a non-independent data set must be searchable by its contents, that is to say that we must be able to pick out all objects which satisfy a given criteria" [GAR 03].

A geographical database is a model that enables us to show data of a spatial nature, using standard relational database technologies [LON 05]. The database essentially proposes the modeling of entities as objects which may be grouped into classes, according to the requirements of the given exercise. Let us take the example of data capture (Figure 1.4): per the approach of the geographical database, this entity becomes an object which is integrated within the wide class, "data capture work". Moreover, a data capture exercise is made up of several types of engineering projects (such as wells, drilling, foggara, etc.). Drilling may therefore also be processed as a class comprising "log", "pumping trial", "equipment", "stratigraphic", etc.

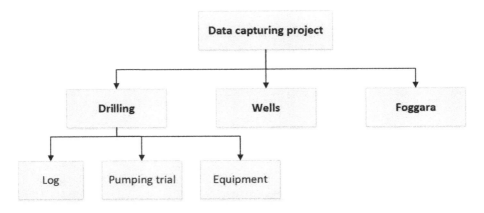

Figure 1.4. *Example of a relational–object database*

The architecture of a geographical database essentially regards a suite of simple yet vital database concepts. The system of database management (DMS) offers a simple, formal model enabling us to archive data in tables. Users have a tendency to think that the DMS is a system naturally open to the simplicity and flexibility of the generic relational data model, enabling them to take charge of a wide number of applications. According to [ZEI 00], the main concepts of a DMS are the following elements (Figure 1.5):

– data is organized in the form of tables;

– tables contain records (in rows);

– the tables are structured in columns;

– each column has a given number or letter (for example: whole number, decimal number, character, date, or others);

– relationships are used to associate rows of one table with those of another table. This operation is carried out thanks to a common column in each table, often referred to as a primary key and foreign key;

– there are rules of integrity for data sets based on tables. For example, each line always shares the same columns, a sphere gives the list of values or value ranges valid for a given column and other such rules;

– the SQL language is made up of a series of functions and operators enabling the carrying out of operations regarding tables and their data; and

– the SQL operators are designed to function with generic relational data (whole numbers, decimal numbers and characters).

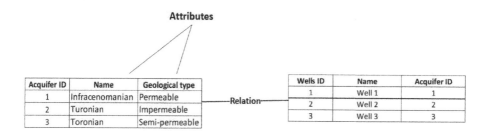

Attributes

Acquifer ID	Name	Geological type
1	Infracenomanian	Permeable
2	Turonian	Impermeable
3	Toronian	Semi-permeable

—Relation—

Wells ID	Name	Acquifer ID
1	Well 1	1
2	Well 2	2
3	Well 3	3

Acquifer class

Wells class

Figure 1.5. *Principle of a spatial database*

So, to model a database effectively, a high number of researchers and engineers adopt a proposal put forward by an organization for American standardization, known as ANSI (American National Standards Institute). This body proposes a three-tier approach to produce a database design: the conceptual level, the external level and the internal level (Figure 1.6).

At the conceptual level: the conceptual data model has as its main objective the modeling of geographical data to take on by means of a database. This model is independent of all constraints of software and hardware, but it prefigures the diagram for the geographical database which must be implemented [DON 06].

External level: at this level, each user has a description of collected data, known as the external diagram [GAR 03]. According to their given needs and prerogatives, they may have a different view of data stored within a DB. This level of abstraction is labeled the external level. This is the highest level of abstraction [BAL 07, ROY 09].

Internal level: this level corresponds to the storage structure supporting data. The definition of the internal diagram necessitates the prior choice of a DMS. It therefore enables you to describe data in such a way that they are stored within the machine.

Several formalisms have been proposed to produce a conceptual data model: E/R (entity/relationship) and UML (Unified Modeling Language) are the best known formalisms. As part of the framework of implementing a water data system, you must opt for the UML formalism. The conceptual diagrams are shown by class diagrams so as to avoid redundancies and inconsistencies in databases being designed. In addition, this formalism is an excellent means of communication, thus contributing to the exchange of information between experts in various fields who contribute to the development of the model.

The UML formalism has already been adopted by some researchers developing environmental databases [GUA 03, MIS 11, WOJ 08]. It is therefore adaptable to the issues of water resource management.

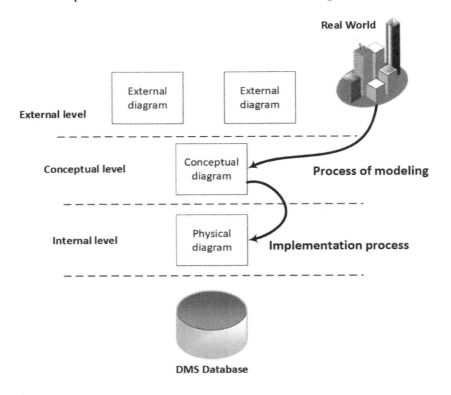

Figure 1.6. *The levels of representation within a DMS according to the ANSI standard*

1.1.3. *Quality of spatial databases*

Within the field of the integrated management of water resources, data comes from various data sources (the Hydraulic Basin Agency, the Ministry for Water and the Environment, OpenStreetMap, OneGeology, etc.). This diversity of data and its sources forces us to take into consideration the problem of quality, the impact of which has a direct influence on decision-making. That being said, it is important to study the data quality, in particular its geometric accuracy, its exhaustiveness, its semantic accuracy and its adequacy for the needs of future users.

The aim of the description of geographical data quality is to facilitate the comparison and the selection of all geographical data best adapted to user needs and demands (Figure 1.7). Complete descriptions of the quality of a data set encourage the sharing, exchange and interoperability of geographical data. Information around the quality of geographical data enables a data producer to assess to what extent a data set responds to the criteria defined within its product specification and helps users assess the product capacity to meet the demands of their particular application (Figure 1.8).

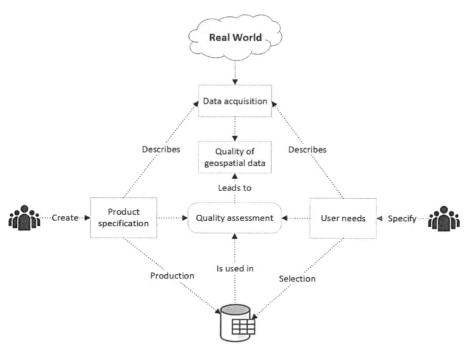

Figure 1.7. *Principle assessment of spatial data quality*

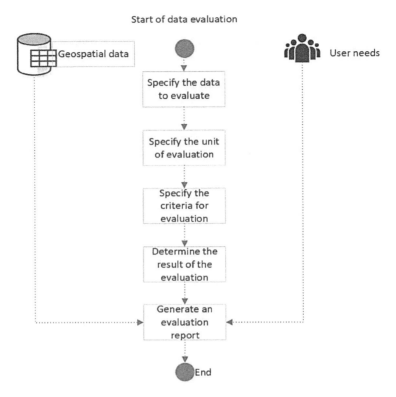

Figure 1.8. *Procedure for evaluating the quality of geographical data*

Quality is measured through criteria (geometric accuracy, exhaustiveness, semantic accuracy, logical coherence, the real world, etc.) and is defined within the ISO 19113 standards with the aid of methods explained within the ISO 19114 standard as well as with specific measures defined within the ISO 19138 standard.

Seven criteria, both quantitative and qualitative, enable the definition of such quality (Figure 1.9):

– semantic accuracy quantifies errors revolving around alphanumeric data;

– exhaustiveness is defined as the presence and absence of geographical entities, their attributes and their relationships. It should be described by the applicable data quality elements applicable from the following list:

commission (excess data present within a data set) and omission (data absent from a data set);

– logical coherence is defined as the degree of adherence to logical rules concerning data structure, attributes and the relationships between given data;

– geometric accuracy enables the supply of data around the gaps between actual locations of a geographical entity and locations described within databases;

– the current data provide information on production dates and on spatial data updates;

– genealogy retraces the history of the data set from its creation or its acquisition by describing the various uses of the data; and

– the specific quality is a "personalized criteria" which enables the user to define their own quality criteria if the previous ones do not meet their expectations.

These criteria enable a user to assess the quality of a given data set and the adequacy of this in relation to their needs. Such criteria enable the producer to assess the adequacy of the data set for the specification and to identify stages which may potentially pose a problem during data capture.

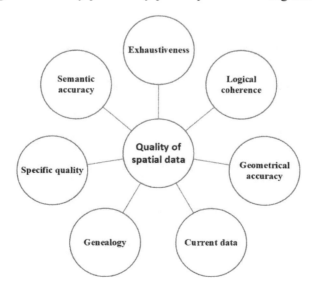

Figure 1.9. *The seven criteria of geographical data quality*

1.1.4. *GIS and water data*

The spatial reference information system, as an autonomous tool or when coupled with various water management models (MODFLOW, HEC-RAS, etc.), may provide solutions to the issue of water management. A hydrological model generally necessitates detailed spatial data owing to the complexity of the natural environment. Previously, the process of modeling watersheds was based on paper-format maps to gather information on the nature of soils, hydrogeology and flood-risk areas. This process is not only tedious, but it also demands more work and has a significant margin for error. The use of spatial data to gather information around a given study area is far more feasible as regards technical and economic aspects and demands a smaller workforce. GIS can be used as a platform to present spatial data in an electronic format. This facilitates data access by using GIS office automation or Internet applications which facilitate updating, modeling and statistical and geostatistical analysis. GIS can be used as a data preprocessing technique before inputting it into hydrological models. The integration of hydrological models within GIS systems has enabled the rationalization of data capture and the ability to better interpret the results of given models. As a consequence, GIS is a valuable tool which is often necessary for planning and managing the environment linked to water.

The soil maps, the DEM, topographical maps, the aerial photos and hydrographical network maps can be used to generate input parameters for a given hydrological model. The acquisition of data is the pivot point for all GIS projects surrounding water resources. The GIS data are available from numerous sources. They can be governmental, private, academic, etc. There are numerous platforms that are developed on the web that supply GIS data and are accessible via the Internet.

New data can be created if existing data does not respond to the demands of GIS analysis or modeling. For example, the delimitation of watersheds can be derived from DEM or digitized data from a topographical map. Data can also be created by using a global positioning system (GPS). Soil impermeability can be generated from IKONOS satellite images by using various statistical techniques such as Fisher's linear discriminant analysis.

In a generic manner, GIS represents a necessary tool within the field of water resources management, as it fulfills the following functions:

– supplying a tool to homogenize data coming from various sources (DWG, SHP, DAT);

– improving the visualization, management, processing and analysis of data;

– improving the understanding of interactions between water and the natural environment;

– enabling the visualization of 3D data;

– managing statistical, geostatistical and digital modeling; and

– enabling the extraction of hydrological data from satellite and photo-aerial images.

1.2. Spatial data infrastructure

All bodies, institutions, companies or groups of individuals, collect, process and produce a significant quantity of "data", the majority of which does not hold particular importance. On the other hand, other data enables better understanding of the given body itself and its environment. This category of data is labeled "information" [LAP 12]. Information is thus a collection of organized data which provides added value in relation to the data itself.

Amongst this information produced, spatial information has a strategic and essential significance, taking an increasingly significant place with the widespread use of various tools: GPS, satellites, sensors, etc. and the multiplication of their applications (Google Earth, OpenStreetMap, etc.) facilitate the integration of spatial data within organizations. This positive observation will encourage the multiplication of geospatial databases according to their uses and geographical areas. These databases, when they are not coordinated, complicate the identification and exploitation tasks for this highly important data.

In 1992, during the Earth Summit in Rio de Janeiro in Brazil, Agenda 21 established amongst its resolutions, the significance of geospatial data for all decision-making processes, as regards environmental preservation and protection. This places the creation, organization and access to spatial information amongst the priorities of all decision-makers. In 1993, the term "geospatial data infrastructure" (GDI) was introduced for the first time by the U.S. National Research Council to designate all technologies, policies and institutional organizations which facilitate the creation, exchange and use of geographical data and annexed information, as part of a data-sharing community.

1.2.1. *Concepts, components and hierarchy*

According to the works of Masser in 2005 [MAS 05] and the Global Spatial Data Infrastructure (GSDI) in 2004 [NEB 04], the purpose of a geospatial data infrastructure (GDI) is to facilitate access to spatial information, through all actions coordinated by organizations, which ensure the implementation of additional and harmonized policies at the national and/or regional level.

To serve this end, a spatial data infrastructure includes all political, organizational, technological elements, standards, financial elements and human resources, enabling you to ensure the increase in value of a wealth of data.

To end up with this vision, the GDI store data, their attributes and their metadata, thus offer possibilities for research, visualization and assessment of their usefulness [GIU 14]. In addition to these basic services, the GDI offer other solutions to a better exploitation of data.

Regarding the works of Masser in 2005 [RAJ 01] and Williamson in 2004 [WIL 04], a spatial data infrastructure includes both technical aspects but also other components of an organizational and political nature, as a GDI cannot be effective without a government authority coordinating it.

Publications in the field [NEB 04, FOR 13] say that a GDI is essentially made up of four elements (Figure 1.10):

– geodata: data and metadata operating and accessible documents;

– technologies: all information technologies which will be used to implement a GDI;

– politics: regulation of exchange and official organizations at the national and/or local level; and

– norms and standards: a protocol reference base agreed for data exchange.

Figure 1.10. *Key components of a GDI*

The various levels of a GDI depend upon the players involved. We can thus define several levels for a GDI: local, provincial, national, regional and global. Figure 1.11 summarizes this vision suggested by [RAJ 02].

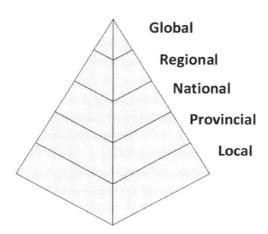

Figure 1.11. *Levels of spatial data infrastructure, from the local to the global levels*

1.2.2. *Interoperability, norms and standards*

1.2.2.1. *Interoperability*

The term interoperability refers to the ability of several systems to exchange heterogeneous data in a cohesive and clear format. As a general rule, respect for standards and norms is necessary to this end [FOR 05]. The International Organization for Standardization [ISO 11a] defines interoperability as being a group of capacities to communicate and carry out programs or to transfer data from diverse functional units in a way that necessitates little or no knowledge relative to the characteristics unique to these units, from the user.

Interoperability within the sphere of geographical technology is a particular case as it processes specific data with various geometric, semantic and syntactic characteristics, and seeks to resolve the problem of format diversity between geographical data.

From a general point of view we can distinguish four levels of interoperability [BRO 16]:

– system level: this represents material or software components necessary for the exchange or manipulation of data and in large part involves aspects relating to the platform, such as operating systems and data exchange protocols (for example, the HTTP protocol);

– semantic level: this is concerned with codification to use for data exchange between IT components. It refers to applications that always interpret data in the same way, so as to provide the desired representation of this data;

– schematic level: responsible for resolving conflicts between the various data diagrams. Within the geographical technology sector, semantic interoperability is essential to close the gap between the terminology used by several GIS software and data sources; and

– human level: this is the human capacity for exchange. Interoperability is often blocked due to a lack of cooperation and willingness to share data.

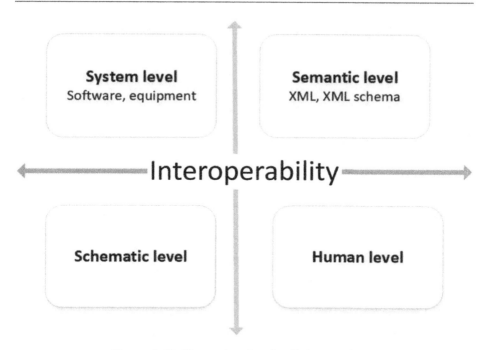

Figure 1.12. *The various levels of interoperability*

1.2.2.2. *Norms and standards*

On a global scale, norms and standards have invaded all manner of public and private sectors so as to provide greater reliability and transparency as well as additional trust between local and international partners, within an ever more globalized world.

This is also the case for geomatic sciences where the advent of the Internet and the development of data sources, including those coming in particular from public institutions and tools for remote sensing of photogrammetry, have overturned a sector which still had a relatively small digital component before the 2000s. The significant growth of this sector for private and public data, which represents a potentially infinite quantity of produced data demands that systems forming the basis of IT management are strengthened, stabilized and that their use is homogenized for thriving work productivity gains, better communication and a greater sharing of geographical data.

Moreover, geographical information has always been a sector of crucial knowledge for military needs and upon which France has historically had a strategic advantage within the agreement between nations.

Next, we will differentiate the concepts of norm and standard. The norm represents a group of technical specifications defined by a group of national or international entities, whilst a standard, which also represents a set of technical specifications, is not validated by a recognized standards body in the form of a new norm for a given sector. Moreover, norms as with standards have the objective of "crystallizing" with a given flexibility a set of best practices which enable, as part of geomatics, the strengthening of the quality of data as well as data-sharing systems.

1.2.2.3. *Norms relating to GDI developments*

According to [FOR 13], norms relating to the development of a GDI can be structured in three categories:

– standards around data contents: these standards essentially relate to the section on data modeling by taking into consideration geometric and semantic characteristics;

– data management standards: the processing of spatial data involves actions such as the discovery of data by means of metadata, the spatial referencing of data, the collection of field data, subjecting data to contributions by entrepreneurs and the juxtaposition of maps developed through images;

– data representation standards: for the visual representation of spatial data with the assistance of sets of symbols for mapped identities.

In the following, we give an overview of the main norms, the use of which is recommended as best practice within the implementation of spatial data infrastructure:

ISO 19103 – Conceptual schema language: this norm enables the supply of basic principles for the development of a conceptual schema in compliance with requirements of spatial information models and aligned with ISO/TC 211 as regards the development of norms. This work also provides directives around the use of the UML to ensure interoperability between models.

ISO 19107 – Spatial schema: this standard allows for the provision of a conceptual schema enabling the description of spatial characteristics of geographical objects. It also defines the spatial operations standard to use for data access, interrogation, management, processing and exchange.

ISO 19108 – Temporal schema: this standard enables the integration of the time dimension within a spatial data infrastructure. It provides a basis to define the attributes of temporal entities, the operations of entity functions and associations, as well as the definition of temporal aspects of metadata, relating to geographical information.

ISO 19109 – Rules for schema application: this standard enables the supply of a metamodel to model the real world. The rules of schema application supply the principles of the abstraction process and the production of application schemas which document a perception of reality. This standard combines parts of the series of standards ISO/TC 211, as it describes how such standards should be used in the development of application schemas. It describes the instancing of global entities of the real world with application schemas, the use of spatial and temporal data types, the inclusion of metadata and quality data, and so on. This standard influences both the definition of application schemas and the development of geographical references in the way that the rules can be considered as requirements of the modeling tool.

ISO 19110 – Methodology for cataloging entities: this standard enables the definition of a metamodel to document the characteristics and properties of the real world. Spatial and temporal dimensions are excluded from the field of application of this standard and are introduced in ISO 19109. This standard favors dissemination, sharing and the use of geographical data, by enabling a better understanding of the contents and the significance of data.

ISO 19111 – System for spatial reference by coordinates: this standard defines the conceptual schema for the description of the geographical localization of a spatial entity through its coordinates. It enables the description of the necessary parameters to define the reference systems for coordinates in one-dimension, two-dimension and three-dimension. This enables the supply of additional descriptive information. This standard also describes information required to convert geographical data of a reference system to coordinates to another system.

ISO 19157 – Quality of geographical data: the objective of this standard is to facilitate the quality monitoring of geographical information. This enables data producers to express to what extent their product responds to the already set out seven qualitative and quantitative criteria, which enable quality definition (semantic accuracy, exhaustiveness, logical coherence, geometric accuracy, relevance, genealogy and specific quality). Moreover, it coherently aids data users in assessing the capacity of a product to meet the demands of its specific application.

ISO 19114 – Quality evaluation: the objective of this standard is to provide guiding principles for assessment procedures for quantitative information around the quality of geographical data, in compliance with the quality principles described within the ISO 19157 standard.

ISO 19115 – Metadata: this standard defines the schema required to describe metadata around spatial information. It provides information around identification, scope, quality, spatial and temporal aspects, the contents, spatial reference, representation, distribution and other geographical digital service data properties.

ISO 19117 – Presentation of geographical information: this standard is intended for IT developers concerned with the approach to use in order to graphically show the use of a geographic data set.

ISO 19119 – Services: this standard provides metadata for the identification of a service that allows access to data or to process it [ABA 12].

ISO 19131 – Data product specifications: this standard provides a conceptual data schema to describe specifications for a geographical data set.

ISO 19136 – Geography Markup Language (GML): the standard's objective is to provide an XML grammar for encoding geographical data. It also enables the description of application schemas as well as the transport of and the methods for storing geographical information.

ISO 19139 – Metadata/implementation of XML schemas: the aim of this standard is to provide an encoding structure using XML technology to describe digital geographical data by defining metadata components, and by establishing a common set of terminologies, definitions and extension procedures.

ISO 19128 – Web map server interface: this standard specifies the behavior of a web service that enables the dynamic publication of map data with a spatial reference from geographical data. It specifies the necessary processes for the recovery of a description around the layers offered by a mapping server, as well as the various processes enabling server interrogation and the display of results. It applies to graphic map previews in image format, but does not apply to vector or coverage data.

1.2.2.4. *Standardization bodies*

1.2.2.4.1. Open Geospatial Consortium (OGC)[1]

According to [BER 07] the OGC is an international non-profit organization founded in 1994, the main mission of which is the development of IT standards within the geospatial community, enabling the interoperability and transparent integration of spatial data, processing software and spatial services. Spatial data and processing comprise geographical information systems (GIS), remote sensing, topography and mapping, navigation, access to spatial databases, sensor networks and other spatial technologies and data sources. Within the consensus process of the OGC, more than 360 governmental, private and cooperative organizations develop, test, document, validate and approve the interface and interoperability problems. The OGC baseline of adopted standards includes such specifications for implementation.

The OGC maintains a class A link with the TC 211 of the ISO, according to which some OGC specifications become ISO standards (Figure 1.13). These two organizations have developed an agreement for cooperation and for the harmonization of their work. This is notably the case for GML specifications which will be the basis of standards ISO 19136 and 19128.

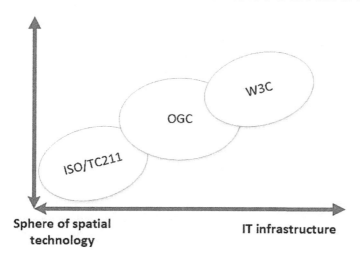

Figure 1.13. *Relationship between OGC and the other standardization bodies for geospatial information*

---

1 opengeospatial.org.

The OGC is in particular responsible for the following standards:

– Catalog Web Service (CSW);

– Geography Markup Language (GML);

– Web Coverage Service (WCS);

– Web Map Service (WMS); and

– Web Feature Service (WFS).

1.2.2.4.2. ISO/TC 211

The TC 211 committee (geographic information/geomatics) was created in 1994 and is connected to the International Standardization Organization (ISO) for which ANSI is the official representative in the USA. The TC 211 established a series of specific ISO standards for geographical information. In view of the significant number of manufacturers within the sphere of geographical information systems and the increasing interest in openness and exchange of data between the various systems, an industrial consortium was established in 1994, known as the Open Geospatial Consortium which currently has 477 businesses, government agencies and universities as member groups. From 1998, OGC and the TC 211 signed a cooperation agreement, and within this framework the OGC adopted a significant number of TC 211 standards. These standards include rules and techniques necessary for the modeling, acquisition, archival storage and visualization of geographical information.

1.2.2.4.3. World Wide Web Consortium (W3C)

W3C is a global organization, created in 1994 with the aim of developing IT standards. Just as with the OGC, the W3C does not put forward norms, but rather industrial standards. The W3C is in particular responsible for the following standards, currently used in the GIS applications developed on a web platform:

– Extensible Markup Language (XML);

– Scalable Vector Graphics (SVG);

– Cascading Style Sheets (CSS); and

– web services.

In January 2015, W3C and OGC announced the start of a partnership within the field of geographical technology with the aim to improve interoperability and the integration of spatial data on the web. So as to establish this collaboration, both bodies each launched a working group with the mission to[2]:

– produce use case documents for publishing spatial data on the web;

– present documents revolving around best practice in order to publish spatial data on the web;

– establish an improved version of time ontology and the network of semantic sensors; and

– work on a new approach for the publication of web coverage data.

1.2.2.4.4. Standardization of geographical information in France

In France, the growing need to use quality geographical data as well as the increased cost necessary for its production and use, is leading producers and users to work closely to a greater degree. This is why we are able to note the existence of several players occurring within the development of standards and specifications relating to geographical technology as follows:

– *The National Council for Geographical Information (CNIG)*: this represents the point of contact and the coordinating structure for Inspire in France. It is positioned with the ministry responsible for Sustainable Development, and its Permanent Secretariat relies upon the expertise and the means of the NGI (National Geographic Institute), the state operator as regards geographical and forestry data. It is responsible:

- for provision to the government of geographical data, in particular as regards the coordination of the contributions of affected players and the improvement of links between them;

- for ensuring within this framework the interoperability between databases and facilitating the use and re-use of geographical information; and

- for leading several working groups spread over a number of commissions (geopositioning, toponymy, data and others).

– *The Commission for Spatialized Information for Data Validation (COVADIS)* is a standardization system for geographical data which is used and exchanged within the ministries for the Environment and Agriculture and with external partners, such as local authorities or indeed even public

---

2 https://www.w3.org/2015/01/spatial.html.en.

establishments, in the form of geostandards that the services should apply, in keeping with the Inspire directive.

– *The French Association for Geographical Information (AFIGEO)*, a state-controlled association per the law of 19th July 1901, which was created in 1986. Its purpose is to contribute to the development of the geographical IT sector in France and abroad. It contributes to discussions relating to the definition and implementation of national and European policies on Geographical Information and gathers a large number of scientific and technical players, such as the French Association for Topography or the *Ordre des géomètres experts* (Land surveyors' professional body). In February 2017, the CNIG and the AFIGEO (French association for geographical information) signed a letter of intent aimed at testing, through to July 2017, possible comparisons between the two structures, in view of the acknowledgment of a certain redundancy in their respective mandates.

– *The French Association for Standardization (AFNOR)* in France has the mandate for orientation and coordination for the development of national standards, entrusted by Decree number 2009-697 of 16 June 2009, relating to standardization. It is a member of the CEN (European Committee for Standardization) and the ISO. For this reason, AFNOR is required to confer upon these norms the status of national standard, either through the publication of an identical text, or by ratification, and lastly, by withdrawing national standards which are inconsistent.

1.2.3. *Description of standards*

A GDI is essentially based on software components called web services. They are able to be used in any language and on any platform, but use the same XML language to communicate with each other through messages. The XML language is used to describe service interfaces to an equal extent as for coding the messages exchanged. Communication may be established through web protocol standards [KOL 03].

According to [POR 08], a web service can therefore be considered as a functionality implemented by the provider of a remote server, accessible to a client tool via the web, without human intervention (automatization) and whatever the technology used (interoperability).

Within the sphere of geospatial technology, the OGC offers several web standards for the dissemination and processing of geographical data

(Figure 1.14). The purpose of these standards is to make GISs interoperable between each other and to produce complex and open systems and data services, favoring content and services accessible to all and usable by all types of application. Moreover, the WMS (Web Mapping Service) and WFS (Web Feature Services), more particularly, enable the service of individual spatial vector entities in a transactional way (selection, insertion, updates and deletion operations). The vector data transmitted are shown in a GML format, which specifies an XML encoding of geometric primitives such as points, lines and polygons. WFS thus enable the handling of geometric data, which is the source of maps [DUB 07].

Using this logic, a body makes its data accessible via the web with the aid of a software server complying with the standards of the OGC (WMS, WFS, etc.), which enables a user of client software, adhering to these standards to visualize, and indeed to handle, the data as if it came into their hands at work [POR 08].

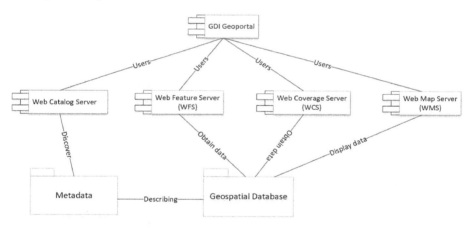

Figure 1.14. *The essential components of a GDI from a technological viewpoint*

1.2.3.1. *Web Map Service*

The WMS specification describes an interface in which georeferenced maps can be made available [ANN 05]. Data are visualized in image format (gif, png, jpeg and svg).

The service is made up of three compulsory processes (GetMap, GetFeatureInfo and GetCapabilities) and four optional processes (GetLegend

Graphic, GetStyle and PutStyle and DescribeLayer) to send requests to the server and obtain information (Figure 1.15).

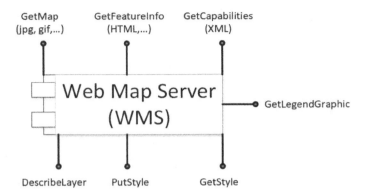

Figure 1.15. *The main processes supported by a WMS*

1.2.3.1.1. GetCapabilities

Thanks to this request, the client interrogates the various servers on the types of maps that they can generate (list of layers, image formats produced, projection systems, etc.).

During a GetCapabilities request, the WMS server must respond to this request by presenting the list and the properties (style, layers, etc.) of interfaces for which the WMS server has responsibility. For example, if a client is looking for data for a base layer for the given catchment area, they can interrogate the WMS by the means of the interface for the extraction of capacities (Figure 1.16). The spatial extent, as well as the keyword and interface projection parameters, prove to be useful for the research of spatial data as a means of service registers and other infrastructures of exploration [GEO 04].

When the WMS server is used, the response for the GetCapabilities request will be in the form of an XML containing a metadata service, according to the given XML schema. This response enables the determination of the type of available data, the possibility of researching information and lastly to define one or several projection systems according to the given user request.

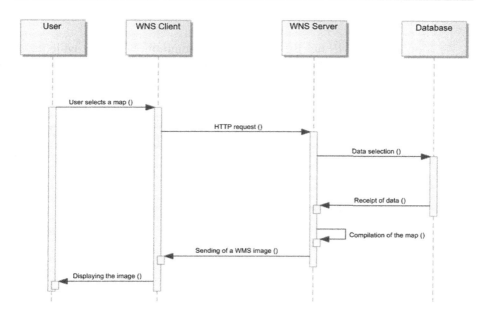

Figure 1.16. *Basic principles for a WMS operation*

Example of a GetCapablities request:

```
http://localhost:8080/Geoserver/services?request=GetCapabilit
ies&version=1.1.1&service=WMS
```

```
1  <?xml version="1.0" encoding="UTF-8"?>
2  <!DOCTYPE WMT_MS_Capabilities SYSTEM "http://localhost:8080/geoserver/schemas/wms/1.
3  <WMT_MS_Capabilities version="1.1.1" updateSequence="152">
4    <Service>
5      <Name>OGC:WMS</Name>
6      <Title>GeoServer Web Map Service</Title>
7      <Abstract>A compliant implementation of WMS plus most of the SLD extension (dyna
8      <KeywordList>
9        <Keyword>WFS</Keyword>
10       <Keyword>WMS</Keyword>
11       <Keyword>GEOSERVER</Keyword>
12     </KeywordList>
13     <OnlineResource xmlns:xlink="http://www.w3.org/1999/xlink" xlink:type="simple" x
14     <ContactInformation>
15       <ContactPersonPrimary>
```

Figure 1.17. *An extract from a GetCapabilities XML response format.*
For a color version of this figure, see www.iste.co.uk/jarar/spatial.zip

1.2.3.1.2. GetMap

As soon as the WMS has the necessary information on the available data of each server, the client can choose the desired layers for the various servers and send a GetMap request to the affected servers, so as to send the desired images on the same map coming from various servers.

When a WMS client launches a GetMap request to a WMS server, the WMS server returns a response (matrix or vector) to the user based upon the following parameters:

– VERSION: version of WMS used;

– REQUEST: type of request (in this case it is the GetMap request);

– SRS: spatial reference system expressed by the EPSG code (European Petroleum Survey Group);

– BBOX (bounding box): desired geographical extent;

– WIDTH, HEIGHT and FORMAT: format of map generated (GIF, JPEG, SVG and others).

The WMS uses the HTTP protocol using the POST or GET method. A WMS delivers purely graphic images in a custom image format such as PNG (portal network graphics), JPEG (joint pictures expert group) or GIF (graphics interchange format). The production of maps is also envisaged in the SVG format.

Example of a GetMap request:

```
http://localhost:8080/geoserver/wms?
request=GetMap
&service=WMS
&version=1.1.1
&layers=hydro
&styles=hydrography
&srs=EPSG%3A4326
&bbox=-5.95,30.6, -3.44,32.6
&width=780
&height=330
&format=image%2Fpng
```

Figure 1.18. *Result of a GetMap request*

1.2.3.1.3. GetFeatureInfo

Through this request, the client can obtain factual information relating to objects. The response will be in the form of an XML document which contains a metadata service, per the XML schema. This response enables the consumer to obtain information in the form of text or attributes about a given region or location. It often happens that this request returns the attributes associated with the graphic component, such as the stratigraphic properties for a given sinking exercise or hydrogeological data. This characteristic is optional within a WMS server.

Example of a GetFeatureInfo request:

```
http://localhost:8080/geoserver/wms?
request=GetFeatureInfo
&service=WMS
&version=1.1.1
&layers=hydro
&srs=EPSG%3A4326
&format=image%2Fpng
&bbox=-5.95,30.6, -3.44,32.6
&width=780
&height=330
&query_layers=topp%3Astates
&info_format=text%2Fhtml
&feature_count=50
&x=-7
&y=31
&exceptions=application%2Fvnd.ogc.se_xml
```

1.2.3.1.4. Definition of SLDs (style layer descriptor)

The SLD provides the possibility to determine the graphical representation of data returned by a WMS server. For example, although the layer of the hydrographical network on a WMS server shows, by default, the hydrographical entities in black, the integration of an SLD into the server enables the WMS client to request that hydrographical entities are, for example, shown in blue.

The SLD enables the monitoring of visualization parameters on the map (Figure 1.19). The SLD offers the WMS client the possibility to determine the type of symbol and the visualization geometry.

For WMS servers using the SLD, there are two specification methods: either the server side or the client side. The specification for the client side requires of the client that they integrate, within their GetMap request, a list of layers recognized by the server, and a corresponding list of styles to use for each layer. The specification of the client side is far more powerful: it is an XML document integrated into the GetMap request, which is sent to the WMS server, and which contains the list of layer styles. The power of this characteristic lies in the fact that the XML document can contain layers and styles determined by the user [GEO 04].

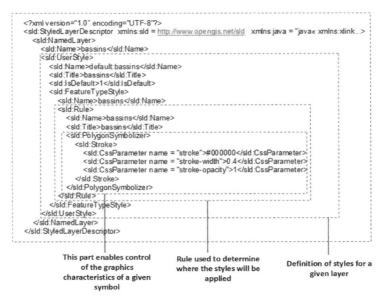

Figure 1.19. *An example of an SLD document*

1.2.3.1.5. Web Map Context

A Web Map Context document represents an additional OGC specification to the WMS specification. The user can save map properties (catchment areas, scale and other factors) in the form of an XML document. The real advantage of context documents lies in the fact that preferred maps may be both saved and activated.

The context documents of the web map are specified from the client side. They are not sent to the server, and the client must interpret and translate them through requests.

A context document is made up of several elements (Figure 1.20):

– geographical extent of the area to display (bounding box);

– geometry of the display;

– data on the individual who has created the map context document; and

– an ordered list of layers indicating the URL server address, the layer name, information stating whether each layer can or cannot be interrogated, as well as data regarding styles.

```
 1 <?xml version="1.0" encoding="utf-8" ?>
 2 <ViewContext xmlns="http://www.opengis.net/context" version="1.1.0">
 3 <General>
 4       <Window width="1280" height="546"/>
 5       <BoundingBox SRS="EPSG:4326" minx="-12.811800347893538" miny="30.4179967378292"
 6       </General>
 7 <LayerList>
 8       <Layer queryable="false" hidden="false">
 9           <Server service="OGC:WMS" version="1.1.1">
10           <OnlineResource xlink:type="simple" xmlns:xlink="http://www.w3.org/1999/xl:
11           </Server>
12               <Name>MONDE_ONEGEOLOGY_PROJECT</Name>
13               <Title>MONDE_ONEGEOLOGY_PROJECT</Title>
14               <FormatList><Format current="true">image/png</Format>
15               </FormatList>
16               <StyleList><Style current="true">
17                   <Name>
18                   </Name>
19                   <Title>Default</Title>
```

Figure 1.20. *Extract of a WMC file. For a color version of this figure, see www.iste.co.uk/jarar/spatial.zip*

1.2.3.2. *Web Feature Service*

The WFS specification enables the transfer of geographical objects. The mechanism for the exchange of GML data for the OGC is used as a basis for

WFS specifications. In this way, geographical data and their corresponding schema can be coded and transferred into XML/GML [ANN 05] (Figure 1.22).

We are able to differentiate Web Feature Services according to two types.

– *Basic WFS*: a basic WFS contains the operations GetCapabilities, DescribeFeatureType and GetFeature (Figure 1.21).

– *WFS Transaction*: this supports all basic WFS and, in addition, the operations Transaction and LockFeature (optional).

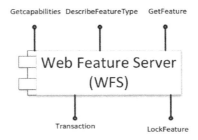

Figure 1.21. *Operations supported by a WFS*

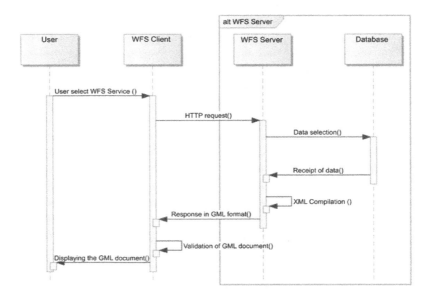

Figure 1.22. *Basic principles for the WMS request*

1.2.3.2.1. GetCapabilities

At the WFS level, the server must respond to a GetCapabilities request, so as to present a list indicating which operations and types of entities it is responsible for [GEO 04]. The GetCapabilities request must respond through an XML document, in accordance with the schema, for response capacities. This document is made up of the following sections:

– service identification: enables the provision of information around the service itself (version, user rights, etc.);

– provider service: provides metadata for the organization that created the WFS service;

– metadata operation: enables the supply of information around operations used by the WFS service;

– FeatureType list: enables the definition of the list of types of entities that are available through the Web Map Service. In addition, this section contains information around the SRS (spatial reference system) forecast system, the geographical limit of the study area, etc.;

– ServesGMLObjectType list: defines the list of GML type objects, which are available through the WFS service which supports the operations of GetGMLObject;

– SupportGMLType list: defines the list of the GML objects that the WFS is capable of using directly, or by means of the GML schema.

1.2.3.2.2. DescribeFeatureType

The function DescribeFeatureType enables the generation of a descriptive schema for entity types used by the WFS service. The schema allows one to define how the WFS encodes the entities upon input (via the requests Insert and Update) and how the entities will be generated as outputs (via the response GetFeature and GetGmlObject).

Example of a DescribeFeatureType request:

```
http://example.com/geoserver/wfs?
service=wfs
&version=2.0.0
&request=DescribeFeatureType
&typeNames=namespace:featuretype
```

1.2.3.2.3. GetFeature

A WFS should be able to deliver objects of entity types (feature). In addition, it must acknowledge which properties must be provided and it must be able to process spatial and non-spatial selections.

The GetFeature request enables extraction from the server of the contents for a single entity (Query – Figure 1.23). The response is an XML document which must comply with the XML schema.

Example of a GetFeature request:

```
http://localhost:8080/deegreewfs/services?
service=WFS
&version=1.1.0
&request=GetFeature
&typename=app:sinking
&NAMESPACE=xmlns%28app=http://www.deegree.org/app%29
```

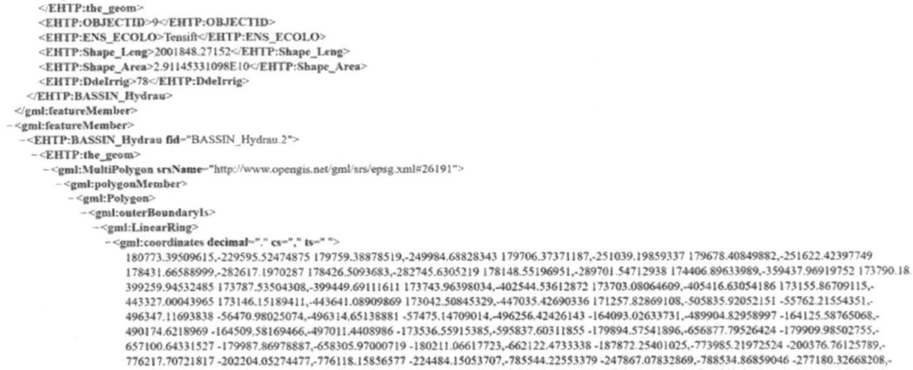


```
      </EHTP:the_geom>
      <EHTP:OBJECTID>9</EHTP:OBJECTID>
      <EHTP:ENS_ECOLO>Tensift</EHTP:ENS_ECOLO>
      <EHTP:Shape_Leng>2001848.27152</EHTP:Shape_Leng>
      <EHTP:Shape_Area>2.91145331098E10</EHTP:Shape_Area>
      <EHTP:DdeIrrig>78</EHTP:DdeIrrig>
    </EHTP:BASSIN_Hydrau>
  </gml:featureMember>
- <gml:featureMember>
  - <EHTP:BASSIN_Hydrau fid="BASSIN_Hydrau.2">
    - <EHTP:the_geom>
      - <gml:MultiPolygon srsName="http://www.opengis.net/gml/srs/epsg.xml#26191">
        - <gml:polygonMember>
          - <gml:Polygon>
            - <gml:outerBoundaryIs>
              - <gml:LinearRing>
                - <gml:coordinates decimal="." cs="," ts=" ">
                  180773.39509615,-229595.52474875 179759.38878519,-249984.68828343 179706.37371187,-251039.19859337 179678.40849882,-251622.42397749
                  178431.66588999,-282617.1970287 178426.5093683,-282745.6305219 178148.55196951,-289701.54712938 174406.89633989,-359437.96919752 173790.18
                  399259.94532485 173787.53504308,-399449.69111611 173743.96398034,-402544.53612872 173703.08064609,-405416.63054186 173155.86709115,-
                  443327.00043965 173146.15189411,-443641.08909869 173042.50845329,-447035.42690336 171257.82869108,-505835.92052151 -55762.21554351,-
                  496347.11693838 -56470.98025074,-496314.65138881 -57475.14709014,-496256.42426143 -164093.02633731,-489904.82958997 -164125.58765068,-
                  490174.6218969 -164509.58169466,-497011.4408986 -173536.55915385,-595837.60311855 -179894.57541896,-656877.79526424 -179909.98502755,-
                  657100.64331527 -179987.86978887,-658305.97000719 -180211.06617723,-662122.4733338 -187872.25401025,-773985.21972524 -200376.76125789,-
                  776217.70721817 -202204.05274477,-776118.15856577 -224484.15053707,-785544.22553379 -247867.07832869,-788534.86859046 -277180.32668208,-
                  804756.24620378 -279712.17949425,-806151.33299049 -280856.18065684,-806784.84550029 -292966.40127028,-813802.55639316 -305458.95593815,-
```


Figure 1.23. *Example of a GetFeature response. For a color version of this figure, see www.iste.co.uk/jarar/spatial.zip*

1.2.3.2.4. Interaction between WFS and the WMS

A WFS operates in direct collaboration with a WMS. The figure below enables us to further illustrate the interaction between these two web services, as well as the sequence of internal actions activated by a GetMap request from WMS.

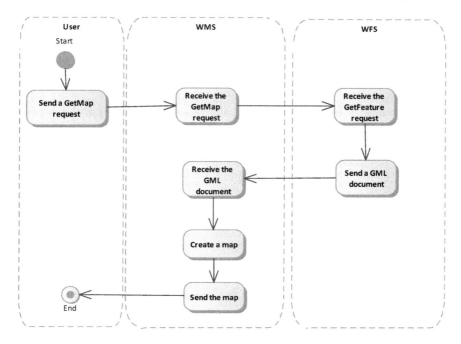

Figure 1.24. *Flow diagram of UML activity for a WMS request*

1.2.3.3. *Web Coverage Service (WCS)*

This service proposes a standard for the behavior of a web service in the production of georeferenced matrix data. It describes the structure of the request to send to the server and how the server returns the response to produce maps.

The WCS servers take responsibility for the following requests: GetCapabilities, DescribeCoverage and GetCoverage.

Figure 1.25. *Operations supported by a WCS*

1.2.3.3.1. GetCapabilities

A WCS server must respond to the request on the capacities through an XML document complying with the XML schema of capacities, defined by the specification of the WCS interface. The capacities must indicate the WCS operations and the operations unique to the supplier which the server takes responsibility for, details around the types of data which the services are responsible for (including interface specifications) and the constraints of server access. Lastly, the server must return links to the external catalogs, which contain and describe metadata for coverage used by the service.

1.2.3.3.2. DescribeCoverage

This operation enables the recovery of a complete description of one or more coverages available through the intervention of the WCS server. The response must take the form of an XML document modeled on the XML schema defined in the specification. For the client, it is a way of finding the coverages available on the server and, for each coverage, recovering the spatial extent, the reference system, the accepted formats and the types of information that contain coverage.

1.2.3.3.3. GetCoverage

A client can recover complete coverage, a coverage limited by a given geographical region or a time interval, or a subset of coverage types. This can feel like a capacity for "uploading data", whilst the user obtains the data (such as GeoTIFF, shapefile, etc.) in place of the GML or a given image.

1.2.3.4. *Catalog Service for the Web (CSW)*

The CSW is a standard enabling the publishing, managing and cataloging of metadata, as well as the interaction with other geographical data registers, remotely via the web.

The notion of spatialized resources can correspond to these mapping data and OGC web services, but also spatial systems of reference or sensors (for example SensorML). Within the framework of this document, the resource is limited to mapping data and web services defined by the OGC (WMS, WFS, WMTS, WCS and others).

The main operations for CSW are:

– GetCapablities: enables the return of information around the capacity of a service;

– DescribeRecord: enables a client to discover the elements of the information model which the target catalog service takes responsibility for;

– GetRecordByID: recovers the representation by default of catalog metadata recordings, by using their identifier;

– GetRecords: enables the interrogation of the recording of catalog metadata specifying a request in the OGC filters or CQL (Cassandra Query Language) languages; and

– transactions: define an interface to create, modify and/or delete catalog recordings.

1.2.3.5. Web Processing Service (WPS)

The WPS essentially serves as a means to effect geoprocessing operations. The processing operations are highly varied. They can range from simple processing (calculating the buffer zone for drilling) to the production of an interpolation map through geostatistic methods or the simulation of the propagation of a flood or a tsunami. The WPS therefore has as its main objective extending the capacities of a GDI by proposing functionalities of processing and spatial analyses close to GIS office automation.

A WPS server enables you to take responsibility for the following operations:

– GetCapabilities: enables one to describe the information around the service and lists the operations and the available methods;

– DescribeProcess: enables the description of the procedures of a geoprocessing service; and

– Execute: enables the launch of the execution of a procedure and the production of a corresponding outcome.

1.2.3.6. SOS Web Service

The Sensor Observation Service (SOS) is a service able to recover data from sensors (meteorological stations, piezometer and others). Whether these are *in situ* sensors (for example water levels) or dynamic sensors (for example satellite images), the SOS enables operators to take responsibility for several operations including, in particular:

– GetObservation: to provide an interface to interrogate, filter and recover data from sensors in real time; and

– DescribeSensor: for the provision of information in the sensors.

1.2.3.7. *Geography Markup Language (GML)*

The GML is an OGC specification based on XML technology. It describes all of the characteristics of geographical entities from one-dimensional to four-dimensional. It also serves as a modeling language as well as a format for open and interoperable exchange over the Internet. It is mainly used by a WFS to send geographical entities between servers and clients. GML includes both the definition of spatial and non-spatial properties of geographical elements (Figure 1.26) [MIG 12]. Several application schemas rely on GML to describe and exchange data within specific fields [BEA 12].

```
 1 <?xml version="1.0" encoding="utf-8" ?>
 2 <ogr:FeatureCollection
 3      xmlns:xsi="http://www.w3.org/2001/XMLSchema-instance"
 4      xsi:schemaLocation="http://ogr.maptools.org/ hydro.xsd"
 5      xmlns:ogr="http://ogr.maptools.org/"
 6      xmlns:gml="http://www.opengis.net/gml">
 7   <gml:boundedBy>
 8     <gml:Box>
 9       <gml:coord><gml:X>-740339.8493368924</gml:X><gml:Y>-1029481.184773482</gml:Y><
10       <gml:coord><gml:X>912373.9419428306</gml:X><gml:Y>590801.6684063673</gml:Y></g
11     </gml:Box>
12   </gml:boundedBy>
13
14   <gml:featureMember>
15     <ogr:hydro fid="hydro.0">
16       <ogr:geometryProperty><gml:MultiPolygon><gml:polygonMember><gml:Polygon><gml:o
17       <ogr:OBJECTID>9</ogr:OBJECTID>
18       <ogr:ENS_ECOLO>Tensift</ogr:ENS_ECOLO>
19       <ogr:Shape_Leng>2001848.27151999995</ogr:Shape_Leng>
20       <ogr:Shape_Area>29114533109.79999923706</ogr:Shape_Area>
21       <ogr:DdeIrrig>78</ogr:DdeIrrig>
22     </ogr:hydro>
23   </gml:featureMember>
24   <gml:featureMember>
```

Figure 1.26. *Example of a GML file. For a color version of this figure, see www.iste.co.uk/jarar/spatial.zip*

1.2.4. *From several services to an entire spatial data infrastructure*

From a technical point of view, the word "standard" is the key to various components of a GDI and it enables interoperability between various geographical data sources. Currently, the technology of web services makes collaborative work between various fields of application possible. The OGC and the ISO/TC 211 have provided the necessary standards and specifications to implement a GDI.

The creation of a GDI with the web services of the OGC and ISO can be produced in two stages:

Stage 1: development of a cataloging and mapping service

In this first stage, the WMS web services and the CSW are used to enable the interrogation for metadata, accessibility and the visualization of geospatial data (Figure 1.27).

The metadata conforms to ISO 19115 (metadata of geospatial data) and ISO 19119 (metadata service) standards. In using the WFS, the mapping capacities of relational objects of the cataloging service may be used to convert the metadata schemas to ISO 19115.

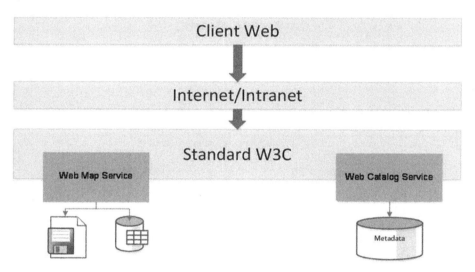

Figure 1.27. *A simple GDI based on a cataloging and mapping service*

Stage 2: direct access to geospatial data

Within this stage (Figure 1.28), the architecture is extended with an integration of WFS and WCS services. This therefore enables greater accessibility, processing and a visualization of data according to access rights appropriate for users, in particular concerning geographical data handling, and also for complex Data Extraction.

These services also enable the creation (digitization) of geospatial data, thanks to the Gazetteer, and the coordinate transformation service.

Since this architecture is the extent of the simplified SDI, it may not be necessary to carry out significant changes to the existing architecture. It suffices to add that new web services, in adapting the client interfaces, offer greater functionalities to the user, according to the access rights attributed thanks to the access control component.

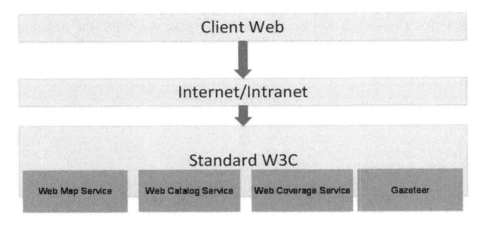

Figure 1.28. *Architecture of developed GDI*

WMS, WFS, WCS and Gazetteer interact in a cooperative manner in the data access control. WMS and Gazetteer support the use of the cataloging service for the provision of spatial recovery mechanisms and/or by using geographical labels. The cataloging service identifies all relevant data and stores information enabling data access through appropriate web services (WMS, WCS and WFS).

All of the services can be decentralized. They can reside on different servers and be maintained by various organizations. This enables decentralized data storage by the parties responsible for specific databases. Moreover, in case of need, the services can be transferred onto other dedicated servers.

1.2.5. *Initiatives*

The set of themes approached by this work necessitates the understanding of different concepts linked, amongst others, to the projects which were developed within the spheres of the geospatial data infrastructures. This part sets out a summary of initiatives revolving around various projects considered relevant.

1.2.5.1. *Infrastructure for spatial information within the European Community (Inspire)*

The purpose of the so-called European directive Inspire is to encourage member countries of the European Commission to develop GDI, favoring environmental protection. In this regard, the main objective of the Inspire directive is to facilitate access to geographical information via a geoportal [ANN 04, EUR 07].

Inspire defines a GDI as a group of geoservices available on a web platform enabling the publication and dissemination, and the sharing of geospatial data. From a technical point of view, INSPIRE is made up of several services (visualization, modification, uploading, etc.) which can be consulted by GIS applications and geoportals via the INSPIRE bus service.

The main objectives of Inspire are [EUR 07]:

– to design GDI within member states of the European Union (EU) that are able to store and manage geospatial data;

– to develop an architecture enabling the combination of spatial data coming from various sources throughout the EU; and

– to develop a platform facilitating the discovery, assessment and use of geospatial data.

1.2.5.2. *Global Earth Observation System of Systems (GEOSS)*

GEOSS is a project which was launched by the intergovernmental Group for Earth Observation (GEO). The objective of this initiative is to implement a global system for the collection of earth observation data. It also enables the interconnection of existing GDIs. GEOSS, through the development of the GEO portal, should serve as a gateway between the producers and the users of geospatial data.

The mechanisms for sharing and for the dissemination of data and information are set out and described within the reference document of the ten-year implementation plan (Secretariat GEO 2005) which data suppliers should agree to and should implement mechanisms favoring interoperability resting upon international given standards. GEOSS is based around existing technologies using satellite services and web technologies.

In addition, project members should fully adhere to the following data sharing principles:

– exchanges of data, metadata and products shared within GEOSS will be open in acknowledging relevant international instruments as well as national policies and legislation;

– all shared data, metadata and products will be available within a minimum period and for a minimum defined cost; and

– all shared data, metadata and products, whether they are free or not, will be used as part of scientific research and education.

1.2.5.3. *Global Spatial Data Infrastructure (GSDI)*

The main mission of the GDSI Association is to promote the international cooperation and collaboration to support infrastructure developments of local spatial data, both national and international, which will enable nations to better approach social, economic and environmental problems urgently [NEB 04]. Its aim is to focus on communication, education, scientific research and partnership efforts to support all of society's needs for access and use of spatial data.

The main aims of GSDI are:

– favoring spatial data infrastructures which support sustainable social, economic and environmental systems, integrated from local to global scale; and

– promoting the use of spatial technologies for the benefit of society.

To support this vision, the GSDI association acts as a platform and offers an extensive choice of publications, conferences, workshops, projects and programs enabling people affected by the given GDI to learn, exchange and share their knowledge and expertise about it.

1.3. Overview of geospatial technologies

In this section, we will provide an overview of the necessary tools for the implementation of a GDI. We have restricted this overview to established projects which an active community is continuing to maintain, and to new projects which appear promising in their capacity to attract a large quantity of users and developers. We have also attempted to provide a general opinion on each GIS category.

1.3.1. *What is GIS software?*

From a technical point of view, the implementation of a spatial data infrastructure necessitates the implementation of a database for the collection and storage of data and metadata, software for data processing and publication, standards and communication protocols and equipment (the latter being computers, servers, cabling, GPS and other related components).

With the appearance of free software, several projects have also come along, amongst them software and databases within the spatial sphere. In 2012, Steiniger and Hunter proposed a model for the technical architecture of a GDI [STE 13]. They identified needs for free and open-source GIS tools to construct IDS tools.

Below is an overview of the history of GIS software and the various categories thereof:

1978	MOSS (Map Overlay and Statistical System), the first vector-oriented GIS software, developed by the U.S. Department of Home Affairs (Carl Reed)
1982	GRASS GIS (http://www.osgeo.org/grass) (Geographical Resources Analysis Support System), the first GIS software to use both vector and raster formats, which were developed initially by the U.S. Army, but only from 1995 did this become open source
1995	UMN MapServer (http://www.osgeo.org/mapserver)
1998	The development of the deegree project (http://deegree.org/) starts by implementing OGC standards
2001	Projects get off the ground: OSSIM (http://www.osgeo.org/ossim), PostGIS (http://www.postgis.org/), open-source GeoNetwork (http://www.osgeo.org/geonetwork), GeoServer (http://geoserver.org/)

2002	Quantum GIS (http://www.osgeo.org/qgis)
2003	gvSIG (http://www.osgeo.org/gvSIG)
2004	uDig (http://udig.refractions.net/)
2005	Beginning of the MapGuide open-source project (http://www.osgeo.org/mapguide)
2006	OpenLayers launched (http://www.osgeo.org/openlayers)
2007	ILWIS – Integrated Land and Water Systems becomes open source (http://wiki.osgeo.org/wiki/ILWIS)

Table 1.1. *Historic development of free and open source GIS tools*

List of some free and open-source software:

GIS Server	North 52°, GeoServer, MapServer, deegree, QGIS Server, MapGuide, PyWPS
Mapping framework	OpenLayers, MapFish, MapBender, GeoTools, GeoExt
Geospatial database	PostGIS, MySQL Spatial, Ingres Geospatial
Spatial analysis tools	PySAL, R Project, Open GeoDA
Mobile GIS	gvSIG Mobile, QGIS for Android, Geopaparazzi
Office GIS	uDig, QGIS, OpenJUMP, MapWindow, SAGA, ILWIS, GRASS
Remote Sensing	OSSIM, GDL, e-foto, Opticks

Table 1.2. *A list of GIS free and open-source software by category*

Several turn-key solutions appeared, enabling the implementation of spatial data infrastructure, using the various components and services. Table 1.3 sets out a summary of these solutions:

	deegree	**GeoNode**	**Georchestra**	**EasySDI**
Launch year	2002	2009	2009	2010
Development of environment	Java	Django (Python Framework)	Java	Java and PHP
Standards and norms	ISO, OGC, INSPIRE	ISO, OGC, INSPIRE	ISO, OGC, INSPIRE	ISO, OGC, INSPIRE
Web services	WFS, WMS, CSW, WPS, WMTS, WCS	WMS, WFS, CSW, WMTS	WFS, WMS, CSW, WPS, WMTS	WMS, WFS, CSW, WMTS

Table 1.3. *List of and operations for free GDIs*

The significance of the use of free software is that it contributes to reducing budget expenditure as regards investment, but especially the openness of the source code. The latter is a major advantage for the personalization and the adaptability of software, even the improvement of its functionalities by developers and administrators responsible for the implementation, supervision and maintenance of GDIs. In addition, what contributes to a better positioning of these tools within the GDIs is the philosophy of free software mentioned by the founder of the free software movement, Richard Stallman [STE 13].

In addition to free software, web services contribute to the implementation of the necessary standards for the operation of GDI. The base architectures for web services (SOA), offer producers and customers the necessary services for the access to and sharing of geospatial data, which improves the interoperability of systems.

Before web services, the GIS applications would exchange their data by utilizing CORBA (Common Object Request Broker Architecture) and DCOM (Distributed Component Object Model). The major disadvantage of these two competitive technologies is that they are not interoperable. Moreover, they are oriented objects or, on the Internet, firewalls prevent the transmission of the Object request, and there is a directory which records them and which permits identification. XML enables both the exchange of data and metadata. From this point of view, it is used both by people and machines.

1.3.2. *Tools*

1.3.2.1. *Mapping servers*

A mapping server is an application responsible for publishing mapping data from several data sources (such as shapefile, raster and others). A mapping server has, in a meaningful way, transformed the data it holds into OGC formats. This section sets out the characteristics of a number of mapping servers.

Geoserver

Geoserver is a server written using Java, the peculiarity of which is that it is designed to be able to use data coming from multiple sources (DMS, shapefile, etc.). Geoserver complies with all forms of standards and norms, whether these are OGC or ISO norms. It implements WFS-T and WCS, supports WPS and uses all of the features offered by HTML5. It offers numerous display formats. Geoserver leads the way in mapping capacities, whether these are possible reprojections, on the cache, operations or regarding the editing. Its documentation proves to be complete, and the project has been in existence for sufficiently long for its bugs to be corrected, and it has a very large user community.

MapServer

MapServer was developed by the University of Minnesota, with NASA's participation. It can be used to publish geospatial data complying with ISO and OGC standards.

From the display angle, MapServer has gaps in reprojection, and around other geometric operations. On the other hand, the excellent cache management and multiple possibilities for data edits are in its favor.

On the support side, MapServer produces average performances: it is well-documented, has examples of varied uses, but is not up to date.

MapGuide

Mapguide comes from an Argus project, modified by Autodesk. In terms of performance, Mapguide is meets the given standards, and proves to be of good quality in terms of styles, formats, and the HTTP protocol. However, Mapguide does not support WPS applications. The Mapguide display capacities prove to be somewhat average, especially in terms of reprojection. The Mapguide support is high quality, although the lack of references and the absence of updates are shortcomings with the application.

QGIS server

QGIS server is the mapping server associated with GIS QGIS, which has been in existence and has been used for around fifteen years. QGIS server meets OGC/ISO standards, implements the SLD styles and manages many display formats. On the other hand, the connection with a WPS can pose a problem, and QGIS server proves to be very poor for the implementation of the WFT-S and WMS applications and in its compatibility with the language HTML5. Fortunately, its flaws do not affect its excellent display capacities.

Although the information is well compiled, it lacks examples, as well as support contacts. There is little return available on this tool, but it is regularly updated and corrected.

MapMint

MapMint is compatible with almost everything: OGC standards, the ISO standards, WFT-S, WMS, WPS applications, except for the HTML5 language. Equally, it does not manage all data formats. This does not alter its exceptional display capacities. This application has full information, numerous contacts and customer views on its use. On the other hand, its flaws are the shortage of examples, and the absence of updates.

deegree

deegree is open-source software designed to build spatial data infrastructures and GIS web applications. deegree offers components for the management of geospatial data, including data access, visualization, discovery and security. It implements OGC standards and those of the technical committee, ISO/TC 211. It recognizes the implementation of the following web services: WMS, WFS, CSW, WCS, WPS and WMTS. The deegree framework has been developed since 2010 OSGeo. Nowadays, deegree is maintained by several organizations and individuals with a large user community throughout the whole world.

1.3.2.2. *Client side APIs*

An API is a programing interface (Application Programing Interface), that is to say, it has code libraries which are already set up and that supply not only functions, but also interfaces.

OpenLayers

OpenLayers is an API with JavaScript functions which appeared in 2006. Its capacities in terms of display (zoom, quality of tiles, flexibility, etc.) have earned it a reputation in the field. OpenLayers also has as advantages its conformity to OGC standards (WFS, WMS, etc.) and the support of various mapping formats, such as GeoJson, KML and GML. OpenLayers can easily be connected to numerous online services such as Google Maps, Bing Maps, and others. OpenLayers enables a multitude of reliable thematic analyses to take place. Data can be obtained in several output formats. From an ergonomic point of view, OpenLayers offers a highly significant speed of execution, coupled with protocols compatible with HTML5, as well as an interface which is clear and well-devised. On the front side, OpenLayers offers comprehensive and clear information accompanied by examples and has a multitude of resource staff. The reputation of OpenLayers has given it a significant amount of feedback and the vitality of the application enables regular updates. Additionally, bugs noted on discussion forums are quickly corrected.

Leaflet

Leaflet is a JavaScript library, which appeared in 2011. It is the library used by OpenStreetMap.

At interface level, this API is as developed as OpenLayers, also having a complete display, numerous formats, compliance with standards, varied output formats, connections to other systems and similar facets. This been said, we lament the less efficient analytical tools.

The ergonomics of the API is of very high quality, as is its general support, but with Leaflet being a more recent library, you will find that fewer projects have used it, so there has naturally been less feedback. The API is regularly updated, but the bugs are not rapidly corrected.

MapGuide

Dating back to 2005, MapGuide is one of oldest APIs on the market. It has a good layer of management, but its length of existence goes against it in terms of the management of external modules, such as the compliance with standards, or the connection to the most current external services. Likewise, its analytical tools are not very high-performance. All the same, it has been able to adapt to the management of data formats, which it knows how to accept or reject without a problem.

Although its interface has no adaptation problems, the speed of execution of requests is a shortcoming as is its low compatibility with HTML5 language. MapGuide is moderately documented, and there are a few examples of its use on the web. MapGuide is not updated and bugs are not always corrected. That being said, user feedback is easy to find.

MapBender

MapBender is an API developed since 2001, as part of the Inspire directive.

We note that its strong points include good layer management, data import/export facilities and multiple connections to other modules. However, MapBender does not support all current data formats. Likewise, the OGC standards are not all respected. Also, its analytical tools are not highly developed.

Even through MapBender is fluid, it is not compatible with the HTML5 language and its interface proves to be obscure which, coupled with its low number of use examples, does not help it to be intuitive. Fortunately, all its information is available.

MapBender updates are rare and bugs persist. A great deal of feedback is available all the same.

Cesium JS

Cesium JS is a library specialized in 3D display. Despite Cesium JS not respecting OGC standards and only accepting a low standard format, its display is high quality and includes interesting zoom possibilities, easily attachable external services and an extensive range of analytical tools.

Cesium JS proves to be particularly high-performance, whether from the point of view of its speed or its interface. It comes with all the necessary information, although it lacks examples. Frequent updates enable bug correction.

MapFish

MapFish is a form of software accessible from the development of architecture which extends the Pylons language and provides tools for spatial data. MapFish brings together the power of OpenLayers and the possibilities of GeoExt, which enables developers to create advanced mapping applications with printing, research and edit functions. OpenLayers and GeoExt guarantee that MapFish conforms to the OGC and respects the standards WMS, WFS, WMC, KML and GML. GeoExt extends a widget interface through ExtJS to understand spatial objects. In addition to implementing the architecture, Python/Pylon, MapFish can also be deployed on the architectures Ruby/Rails or PHP/Symfony.

1.3.2.3. *Cataloging spatial data*

All data which will be integrated into GDI come from various organizations, each using methods and equipment which are unique to it. All data will have their own metadata, which must be both stored and managed.

GeoNetwork

OSGeo (Open Source Geospatial Foundation) defines the GeoNetwork as a cataloging application to manage geographically referenced resources. Geonetwork provides powerful edit and research metadata, as well as a map viewer being included.

Open-source GeoNetwork was developed to connect communities of spatial information and their data by using a modern architecture which is both powerful and low-cost, based on free and open-source (FOSS) software principles, protocols and open and international service standards (ISO/TC 211 and OGC).

M3CAT

M3CAT is a high-performance tool, if you discount its import/export problems in XML and its little developed research capacities. The application turns out to have a poor level of information and is not very sophisticated. It has an obvious lack of updates.

MD-web

MD-web, is a free tool for cataloging and resource localization (data, information and services). It relies on the metadata standards in force (ISO 19115, 19119, 9110 and Dublin Core) and on the transmission (CSW-2 of the OGC) of the geographical information.

One of its originalities is the association of spatial and semantic aspects in the description and the resource research. This is produced by the use of thematic (a dictionary of specialized terms) and spatial (geographical points of interest) repositories, unique to the intended application. It also provides extended functionalities as regards automated capture, the management of these systems of reference (template editor, the spatial basis and the dictionary of specialized terms).

CatMDEdit

CatMDEdit was developed by the University of Zaragoza (Spain). Extremely good technical capacities make it a recognized software. However, its security system is lacking with, in particular, access flaws in the system. The application itself is very well-documented, but there is a shortage of examples. The software is still being developed and regular updates are fixing the bugs in the software.

1.3.2.4. *The spatial DMS*

A spatial DMS is a database management system specialized in the management of spatial data. Its main function is to store a large amount of geospatial data and to enable users to effect optimized requests.

PostgreSQL

Created in 1995, PostgreSQL is a relation and objects database management system (RO DBMS).

The capacities of PostgreSQL enable it to carry out more than 300 requests, to be able to edit, convert formats and efficiently store data. PostgreSQL also proves to be a secure DMS with its functions of identification and role assignment. PostgreSQL is fully documented and regularly updated.

MySQL

MySQL is an open-source DMS under the control of Oracle, the name of which comes from the designer's daughter. It should be in no way envious of PostgreSQL as regards technical ability, although its method of access control performs less well. MySQL has comprehensive support, as well as updates and bug fixes.

SQL server

SQL server was developed by Microsoft. Despite not being designed to manage geographical data, its results are more than reasonable. Its security is of a slightly lower quality but remains average.

Although its results fall slightly short, the documentation relating to SQL server is slightly below the other DMSs. Its updates and bug fixes are also less frequent.

SQLite/Spatialite

Spatialite is a spatial extension of SQLite, a DMS, the peculiarity of which is being directly implemented in the applications. This architectural particularity does influence its capacities slightly in terms of storage, but does not alter its technical performance. That being said, the methods of authentication and role assignment are not perfect.

Although the examples of use of Spatialite are not highly numerous, it is fully documented, and the application is particularly sophisticated.

H2/H2GIS

H2GIS is a spatial extension of H, a DBMS (database management system) written in Java. Being a small application, H2GIS' capacities for storage and editing are far from being extraordinary. However, the other functionalities are satisfactory. The security of the DBMS is properly managed. We note that H2GIS has an average support quality, which might be explained by the application's low level of sophistication.

1.3.3. *Comparative study of current solutions*

1.3.3.1. *Why have we produced this study?*

This comparative study was produced based on available documentation [BRO 17, STE 13] and feedback. It should aid developers in choosing the best components to implement a GDI around water, amongst which are mapping servers, spatial DMS programs, geocataloging systems and client side APIs. Each of these elements will be evaluated according to the criteria and grouped together by category, which will enable their classification, so as to choose the most appropriate applications. Each criterion will receive a weighting according to its capacities, graded thus:

– A: excellent;

– B: good;

– C: average;

– D: there is the existence of general functionality; and

– E: there is a complete absence of functionality.

1.3.3.2. *Evaluation of mapping servers*

Evaluation criteria are grouped together in four categories. We will grade the display, the ergonomics, the support quality and the sophistication of the application.

The mapping servers will be evaluated using four criteria categories: the criteria of conformity to standards, of display, of the documentation available and of the server sophistication.

For standards and norms, the server should conform to the OGC and ISO standard, implement the WFS-T and the WMS, be able to support the SLD and the WPS styles. Being able to use at least five data formats is also necessary.

On the mapping side, reprojection, geometric operations, cache management and data editing will be evaluated.

For support quality, we will assess the application information available, examples of API usage and whether or not resource staff support is available.

The sophistication of the application will be evaluated by means of updating mapping servers, bug fixing and even user feedback.

Standards and norms	Conforms to OGC/ISO standards Implements the WFS-T and WCS Implements the SLD styles Supports a WPS Uses the HTTP protocol (in the modes GET and POST) Supports more than five display formats (Geojson, KML, GML and others)
Display and mapping	Reprojection Supports geometric operations Cache management Designing/editing data
Quality of support	Information Example Mailing list
Sophistication of application	Feedback Bug fixing Updating

Table 1.4. *Evaluation criteria for mapping servers*

According to this comparison, "Geoserver" satisfies all criteria scoring 100%, then as a second option we find the mapping server, deegree, which is limited as regards cache management during the mapping display. That being said, the server MapMint is put at a disadvantage due to the quality of support and the lack of sophistication of the application with its absence of examples, bug fixes and updates.

Summary of evaluation of mapping servers

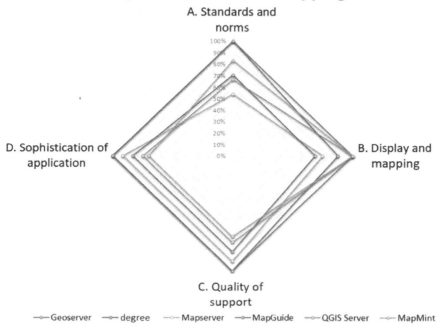

Figure 1.29. *Evaluation of mapping servers. For a color version of this figure, see www.iste.co.uk/jarar/spatial.zip*

1.3.3.3. *Evaluation of client side APIs*

Evaluation criteria are grouped together in four categories. We will grade the display, the ergonomics, the support quality and the sophistication of the application.

For the display category, a grade is given to the display itself (zoom, tiles, etc.), a further for the possibilities of linking the API with external modules. A further criterion will be whether or not the API respects the OGC standards. The API should support at least five data formats, as well as having analytical tools. A final criterion will be the export capacities of the API.

As regards ergonomics, speed will be assessed, as well as the compatibility with the HTML5 language and the facility for using the interface.

For the support quality, we will evaluate the information on the application, the examples for API use and whether or not resource staff are available.

The application sophistication will be assessed through API updates, bug fixes, and even user feedback.

Display	Zoom/section/layer management
	Connection to external services (Google Maps, Bing Maps, OpenStreetMap)
	Use of standards OGC (WFS, WMS, etc.)
	Support of more than five display formats (Geojson, KML, GML, etc.)
	Thematic analysis
	Printing/export of data
Ergonomics and speed	Speed of execution
	HTML5
	Facility for use of the interface
Quality of support	Information
	Example
	Mailing list
Sophistication of application	Feedback
	Bug fixing
	Updating

Table 1.5. *Evaluation criteria for mapping servers*

The most appropriate client side APIs which have the greatest number of functionalities are: OpenLayers, Leaflet and Cesium JS. Amongst these three tools, the most distinguished client side API is OpenLayers, with a score of 100% across all criteria. We then find in second position the API Leaflet. Despite fulfilling a large number of criteria, it has certain limits concerning thematic analysis, feedback and bug fixing. The API Cesium JS has a multitude of limits and offers less functionality than the two APIs OpenLayer and Leaflet. Indeed, Cesium JS has certain limits concerning the implementation of OGC standards (WFS, WMS, etc.), supported formats, examples, the mailing list and feedback.

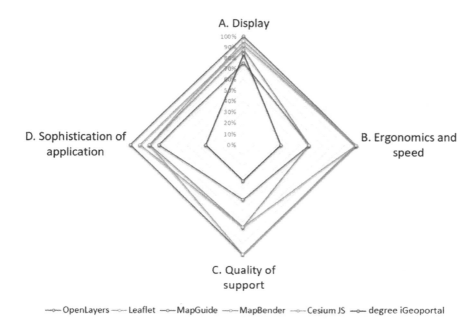

Figure 1.30. *Evaluation of client side APIs. For a color version of this figure, see www.iste.co.uk/jarar/spatial.zip*

1.3.3.4. *Evaluation of spatial DBMS*

The selection criteria for DBMS are the capacity to complete more than 300 different processes, editing functions, the conversion of formats, geometric processes and storage (for the functionalities category), role assignment, the security of authentication, the protection of exchanges, the control of access (for the security section), etc.

The information and the sophistication of applications will also be evaluated according to the same criteria as previously.

Spatial functions	More than 300 processes Editing function Conversion formats Geometric processes Storage/interrogation
Security	Roles Identification and authentication Protection of exchanges Control of access
Quality of support	Information Example Mailing list
Sophistication of application	Feedback Bug fixing Updating

Table 1.6. *Evaluation criteria for DBMS*

According to this summary of geospatial databases, PostgreSQL fulfills the necessary criteria by achieving the best score of 100%. Then MySQL and SQLite, despite their respective rankings, have certain limitations.

MySQL has a limitation as regards the security aspect, since some functions of access control are not implemented for this database. SQLite has a significant number of limitations, particularly regarding security, support and sophistication aspects.

Summary of evaluation of DBMS

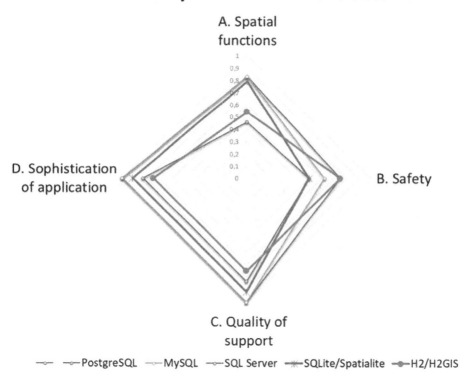

Figure 1.31. *Evaluation of spatial DBMSs. For a color version of this figure, see www.iste.co.uk/jarar/spatial.zip*

1.3.3.5. *Geocatalogs*

The criteria used to assess the various tools which enable the interrogation of cataloging servers for geospatial data are:

Capture function and editing	Standard ISO 19115 XML import/export format Storage of metadata
Publication and research	Multicriteria research Multi-language Interface personalization
Administration and security	Management of roles (user, administrator, validator, etc.) Catalog management Access management
Quality of support	Information Example Mailing list
Sophistication of application	Feedback Bug fixing Updating

Table 1.7. *Criteria for evaluation of a geocatalog system*

According to this summary of the geospatial data catalogs, GeoNetwork fulfills the necessary criteria, with the top score of 100%, whilst deegree CSW and MD-web, despite their classifications, have certain limitations. Indeed, deegree CSW is limited as regards the publication and research category, since the personalization of the interface is delicate for deegree CSW. On the other hand, MD-web has a significant number of limitations, in particular as regards support and sophistication aspects.

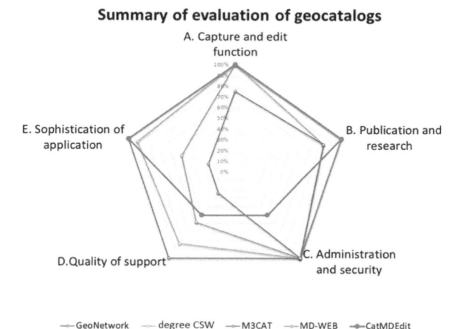

Summary of evaluation of geocatalogs

A. Capture and edit function

E. Sophistication of application

B. Publication and research

D.Quality of support

C. Administration and security

—○— GeoNetwork   —— degree CSW   —○— M3CAT   —○— MD-WEB   —◆— CatMDEdit

Figure 1.32. *Evaluation of geocatalog systems. For a color version of this figure, see www.iste.co.uk/jarar/spatial.zip*

1.4. Conclusion

In closing this chapter, we consider that having gone over the theoretical bases will enable readers to define the general concepts of geospatial technology and their features in the management of water resources. This theoretical framework is subdivided into three sections. The first one is an initiation around geographical information, by stressing the crucial significance of the quality of geospatial data. The second section examines the basic principles of a GDI and states the significance of interoperability as well as its various forms. We also define the norms and standards commonly used in the implementation of a GDI. The third part provides an overview of present geospatial technologies as regards the development of a GDI. A study of benchmarking is covered in detail, enabling future users to better choose between the tools available on the open source market. The theoretical framework as a whole is an efficient method which will assist the reader to better understand the technical aspects and the proposed case studies to come.

2

Technical Framework: Spatial Data Infrastructure for Water

2.1. Introduction

The immeasurable quantity of water data from the collection of different sources should be exploitable and accessible to potential users at different levels. To overcome the barriers limiting optimal data use, [MAS 05] identified a set of needs such as the elimination or reduction of restrictions on data access and availability and the promotion of interoperability between several data sources and different information systems. The set of proposed measures promotes the sharing and exchange of geospatial data at different political and institutional levels. This underlines the undeniable importance of interoperability in the improvement of the quality of use and sharing of water data.

At country level, geospatial data infrastructures (GDI) for water are spaces for the exchange and sharing of water information and data produced by the different operators in the water sector. In fact, they provide a partnership framework between all the operators concerned with this sector, be they public operators (ministries, administrations, agencies, etc.), private interests (businesses, consultancy firms, etc.), researchers and specialists or more broadly the "general public". In this framework, the GDI for water is a social project whose purpose is to offer to decision-makers, experts, scientists and the general public the means to access information on water, reuse it and share it. Hence the importance of geospatial technologies for the development of technical infrastructures for sharing water data and information.

2.2. Water data management

2.2.1. *Water data*

Water exists at the level of our terrestrial globe in significant quantities in three physical phases: liquid, solid and gaseous [ANC 12]. It is also found in the three main categories which are easily accessible to humans: atmosphere, oceans and surface and groundwater. In general, water can easily switch from one state to another and can change from one phase to another in response to its environment. It is a dynamic environment both in space and in time.

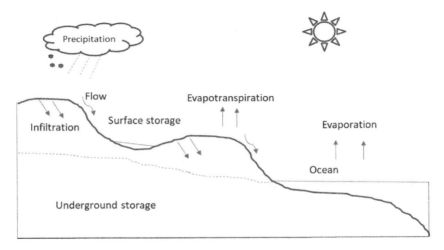

Figure 2.1. *Water cycle*

The field of hydrology focuses on the water cycle as it interacts with the surface of the earth; hydrological data is collected in order to increase our understanding of this interaction (Figure 2.1). Such observations can occur at any point in the hydrological cycle, each employing different techniques for the measurement and estimation of the quality and quantity of water. In this sense, we define several broad categories of water data (Table 2.1).

Water data group	Example of parameters for each group
Climatological data	Temperature, precipitation, evaporation, wind speed
Hydrological data	River flow, river height, flood plain, water speed, physico-chemical quality
Hydrogeological data	Piezometric level, stratigraphic logs, aquifer thickness, permeability, transmissivity
Water quality data	Nutrients (nitrate, phosphorus, etc.), pesticides, pH, turbidity
Complex data	Results of hydrological models, water storage estimates, complex physico-chemical calculations, indicators
Administrative data	The cost, action limit of a hydraulic basin agency

Table 2.1. *Examples of water data parameters*

– *Water observation data*: data representing a set of values collected from hydrological or climatological stations. They are collected by many hydraulic agencies as well as by hydrologists or meteorologists. Three main types of temporal observation data are identified:

- data collected continuously over time using sensors (piezometric level, quality of surface water);

- data collected via sampling sites and later analyzed in a laboratory;

- data collected in the field (stratigraphic log of a well, geophysical data, etc).

– *Spatio-temporal water data*: data collected continuously over time over a geographic range (precipitation, flow, etc.) necessary for the quantitative management of water resource.

– *Spatio-temporal and multidimensional data*: multidimensional, spatio-temporal networks are generally collected or generated for a fixed spatial range (data from meteorological and hydrological modeling).

– *Static geospatial datasets*: static geospatial datasets contain information that makes it possible to characterize the hydrological state of a drainage basin. These statistics cover the following entities: drainage basin limits, waterway hydrography, soil types and digital elevation models (DEMs). Static geospatial datasets can be structured in vector or raster mode.

2.2.2. *Water data sources*

Water data are also distinguished by their source. There are four broad source categories for water data: direct measurement, geo-informatic data (from geographic information system (GIS) processing and remote sensing), estimation from hydrological models and administrative data collection. Each of these categories of water data poses problems of adequacy, accuracy, spatial coverage, temporal frequency and cost.

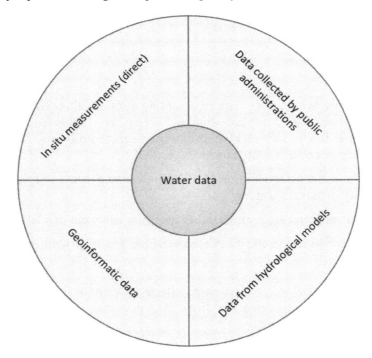

Figure 2.2. *The four categories of water data sources*

2.2.3. *Water data management model*

The implementation of a water data management system is accompanied by a protocol defining the life cycle of water. The latter, with its phases of collection, transmission, updating, archiving and recovery must be adapted to the local context and the dynamics of which it is part.

Data control, verification and validation processes must be integrated as early as possible into the analysis process and the methodology must be adapted to the available human resources.

The time factor is in fact an important point. It is essential to predict in advance the workload related to data management (input, update, etc.) and to include it in the job descriptions of the persons concerned.

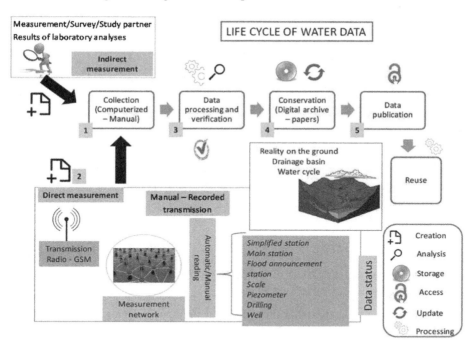

Figure 2.3. *Life cycle of water data, from direct and indirect measurement data to processing for decision-making at the level of the drainage basin*

We can distinguish several steps in the life cycle of water data (Figure 2.3):

– *Data collection*: this can be accomplished in two ways. It can be accomplished directly from measuring stations located within a drainage basin, the reading of which is either automatic via recording memory sensors or manual via visual reading; or it can be accomplished indirectly via analysis laboratories which carry out checks on the quality of the water as part of measurement campaigns or through surveys, polls and studies on water resources, carried out by consultancy firms subcontracted by public bodies.

– *Data transmission*: the transmission of these data to the agency is accomplished either: automatically, via the GSM network or the radio waves of communication devices which accompany measuring stations; or manually, via lists and measurement sheets, transported by agents of the agency or transmitted to the transmission center via walkie-talkies and telephones.

– *Data verification and processing*: in this step, the data are preprocessed in appropriate resolutions and formats. A verification step of measurement quality (especially manual measurements) is carried out by the agency, in order to proceed to the retention stage by classifying and archiving these data.

– *Data archiving*: next, the data are archived for later use. There are two types of archiving: paper archives, which constitute a very important heritage in volume, quality and cost; and digital archives, which are mainly stored at the level of the agency's intranet.

– *Data publication*: after archiving, the data is made available through a water information system.

– *Data reuse*: the last step of this work flow takes the data collected on water and reuses it for something else.

It should be noted that this data management model can be modified as needed.

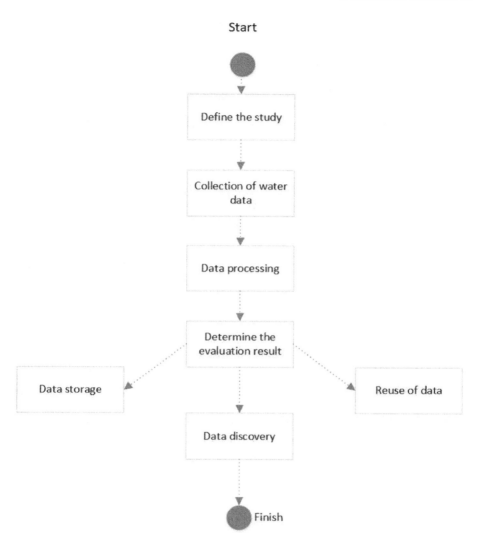

Figure 2.4. *Flow diagram of the data cycle on water*

2.2.4. *Water data interoperability*

Access to and sharing of water data represents a major challenge for users including engineers conducting integrated studies on water resources handling several data sources.

The difficulty in accessing data lies in the unavailability, fragmentation and heterogeneity of the data, which makes it difficult to exploit because of the immense amount of work required to restructure the data in a usable format. This therefore requires a homogenization of different data sources at several levels whose GDI standards are used at each level and which are developed in coordination with the OGC, ISO and professional organizations such as the World Meteorological Organization (WMO) [FIT 16].

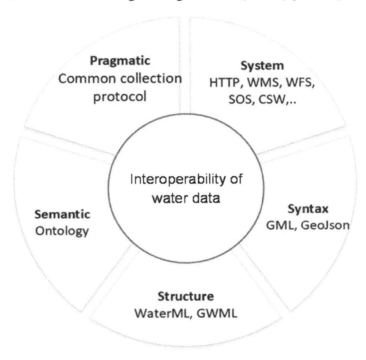

Figure 2.5. *Main levels of water data interoperability*

According to Brodaric *et al.* [BRO 16], the five levels of interoperability are (Figure 2.5):

– the system level concerns the use and deployment of OGC and ISO/TC211 geospatial web services, such as WFS, SOS and WMS, that transmit water data;

– the syntax level involves GML and GeoJson for data coding;

– the structure level concerns essentially the use of related standards and norms and water data codings such as WaterML 2 [TAY 12] and GWML [BRO 18];

– the semantic level refers to the use of standard concepts and associated terms. Terms are usually organized in vocabularies or in code lists and concepts are generally organized in information ontologies;

– the pragmatic level involves essentially the use of common scientific protocols for data processing and collection.

2.3. Establishment of a water information system

Making water data available to the public in a user-friendly and usable format is the ultimate purpose of a water information system (WIS) or of geospatial data infrastructures dedicated to water resources (GDIW). According to the USGS (United States Geological Survey: http://www.usgs.gov/), a water information system is defined as all data collected, stored and disseminated to the general public from a number of information sources connected via a network of computers and data servers [USG 98]. According to EauFrance (http://www.eaufrance.fr/) [EAU 10], a water information system is defined as the system which organizes the production, collection, storage, recovery and dissemination of water data. From a technological point of view, what has encouraged the emergence of these systems is the developmental evolution of three very important concepts: from "open data" and "service-oriented architectures" (SOAs), to the emergence of geospatial data infrastructures (GDIs). The contribution of SOAs is to be able to develop applications that provide a service to other applications, while remaining technologically independent of service suppliers and consumers. For spatially referenced data, the so-called "geographical" web services,created in the early 2000s [POR 08], make it possible to carry out geographical or geomatic processes, for example: providing a map in image format, publishing a catalog of data and metadata and the transformation of coordinates. These geographic web services developed by the OGC are intended to allow greater interoperability between different geographic information systems. Regarding "water web services", the Consortium of Universities for the Advancement of Hydrologic Science (CUAHSI, http://www.cuahsi.org/) has developed since 2006 a suite of web tools and services that allow a better exploitation of water resource data [TAR 11].

2.3.1. *Technical architecture of a GDI for water*

According to Bermudez *et al.* [BER 11], a water information system will have to be based on an SOA architecture allowing publication, discovery and access to the data of which a catalog system represents a mediation service between data providers/producers and users. The main components of this system are (Figure 2.6):

– Data providers include federal, regional, state, local, international and non-governmental organizations, universities, research teams and volunteers.

– Consumers of hydrological data include hydrologists, scientists, engineers, researchers, planners, decision-makers, students and any other party interested in acquiring data on the aquatic environment. Consumers rely on desktop- or web-based client applications to facilitate searches through catalogs accessing data services by analyzing and using data.

– Catalogs support data discovery based on indexed metadata, in the same way that search engines support Internet content discovery. A catalog provides a centralized registry of descriptive metadata and services about data published through these services. Catalogs also provide a search interface allowing data consumers to discover the data that are of interest to them.

– Processing services provide specific features for data operations, such as semantic mediation, transformations, language support, etc., assuming that these operations and their execution can be reasonably factorized and distributed on the Internet.

In addition to these components, we can add other services:

– data services provide an interface to access data from producers and providers of water data;

– research services provide an interface allowing consumers to search for and discover water data. The catalog provides the metadata making it possible to access data using data services.

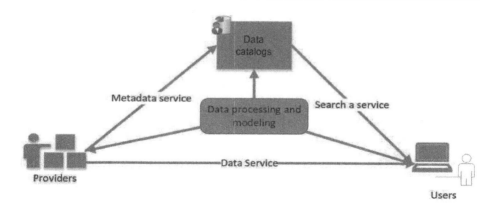

Figure 2.6. *Service-oriented architecture for water (SOA-Water)*

Based on the SOA concept cited above, we can define the global architecture of a GDI for water with the following main components (Figure 2.7):

– A one-stop shop or a geoportal is a user-friendly interface for basic users, experts and decision-makers. It must allow the user to see the data (attributary or chartographic). Classic display and navigation features must be available. The main purpose of this module is to allow to the user to consult the data that they can then download or use via the platform for further use in their own field of competence.

– A cataloging service of data and their metadata to allow better organization and archiving of stored data. This service also facilitates the search and exploration of documents, for example by using thematic thesauri and ontologies. The catalog must allow the user to perform a metadata search and, if they have sufficient rights, to enter metadata records (for partners). The catalog must comply with ISO standards and allow complete interoperability of other metadata catalogs by adhering to the OGC CSW metadata exchange protocol.

– A geoprocessing server which allows the remote exploitation of a rasnge of geographical processing, via tools and appropriate web services.

– A collecting device making it possible to put at the disposal of the hydraulic agencies a notification mechanism of the need to recover partners' data. Data producers under contract are also notified of instructions to send/remove the agreed data. This device also has a feature for the integration and validation of the data received in the system.

– A cartographic atlas aims to publish predetermined atlas maps, allowing the user to assess the coverage by type of data, by size and theme, geographical situation, producer, etc.

– Reporting aims to generate a report on a selection of data (indicators, maps, graphs, etc.).

– A relational geospatial database coordinates the centralization and storage of data.

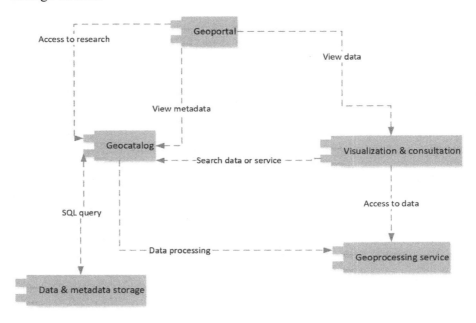

Figure 2.7. *Main components of a GDI on water*

The platform may propose other topics (news, partner directory, documentary space, publication of reports, etc.). The portal for water can also reserve specific spaces, namely:

– A space on dynamics where water indicators which will monitor the state of hydrological systems are grouped together.

– A space dedicated to reports and studies on water.

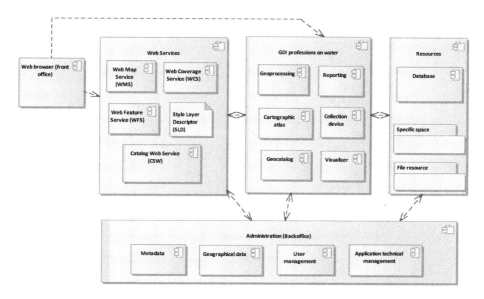

Figure 2.8. *Global architecture of a GDI on water*

2.3.2. *Interaction diagram of main modules*

The geoportal is the favored access point of a GDI for water. However, direct access can be accomplished with the catalog module or with the visualization and consultation module (Figure 2.9).

– the home page allows access to the topics and articles of the site;

– geocataloging makes it possible to search for metadata on an indicator and geographical data to view, display or extract;

– the visualization and consultation module allows interactions with attributary or geographic data (indicators, maps, etc.);

– the cartographic atlas allows interactive consultation of predefined maps;

– the reporting module allows, after selection, the extraction in pdf format of a report on an indicator or piece of geographical information;

– the functional and technical management of the application is ensured by the administrator module;

– the supply of the database can be ensured by the collecting device module using ETL scripts.

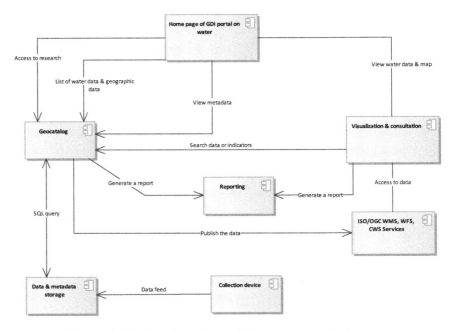

Figure 2.9. *Interaction diagram between the application modules of the platform of a GDI on water*

2.4. International experiences

2.4.1. *At Mediterranean basin level*

Several reports and studies [JOS 11] confirm that the water resources of the Mediterranean region are scarce and very sensitive to many factors: demographic development, increased urbanization, as well as the impacts of climate change and industrial and tourist activities. The Mediterranean region is thus subdivided into three main areas: the North (European) characterized by an abundance of water resources and the South and East which suffer from a lack of resources in the face of a growing demand [ORJ 14]. By 2050, these regions are expected to face a very critical decline in available water resources [MIL 13].

Thus, the strategy for the sustainable development of the Mediterranean region was adopted in 2002 by the 21 Mediterranean countries in addition to the European Community in the framework of the 2001 Barcelona Convention. This strategy has identified seven priority fields of action, including the field on "Integrated management of water demand and resources". Thus, several strategic axes have been developed for this field of action, including the axis: "Information exchange and data collection on water resources," hence the UfM (Union for the Mediterranean) initiative for the Euro-Mediterranean Information System on Know-how in the Water Sector (EMWIS) [SEM 05].

In partnership with operators in the water field at the level of the member countries of the UfM, EMWIS carries out missions and studies for the implementation of national information systems at the level of the member countries of this initiative.

For the member countries of the European Community in the Mediterranean, a European directive for a spatial data infrastructure, "Inspire" [EUR 07], establishes the general rules for a European geospatial data infrastructure, based on the infrastructure of the Member States. Within the framework of the Water Framework Directive (WFD), the members (states) are required to communicate to the European Community information on water resources at basin level. Hence the creation in 2007 of the WIS in Europe, the Water Information System for Europe (WISE) (http://water.europa.eu/). This system is linked to the Inspire European infrastructure and guarantees accessibility to water information at the European level for the public and specialists . By contrast, in the Eastern and Southern Mediterranean countries, the EMWIS system is the only partnership framework for sharing water information at the regional level.

2.4.2. WIS in France

The 1992 version of the Water and Aquatic Environments Act [ONE 10] initiated the National Water Data Network (NWDN) and then, in 2000, France adopted the Water Framework Directive of the European Union, which defines a framework for a community policy withi regard to water. In 2002, France ratified the Aarhus Convention for the right of access to information, public participation in decision-making processes and access to justice in environmental matters. The first work on the WIS project in France dates back to 2003 and it is the 2006 version of the Water Act, which introduced Article L213-2 defining the WIS as the official source of water information in France as coordinated by the NOWAE (National Office of Water and Aquatic Environments), created in 2007. Eaufrance.fr is the single access point for all public water data in France. This web portal groups together links to the 15,000 data producers, the 30 official websites, the 1,200 data sources and the 15 national reference banks on water. Eaufrance.fr represents the access point to a suite of online WIS modules and services at national and regional level (the geoportals of basin agencies). Thus, from this single access point, we can access thematic maps, geoportals, graphs, statistics, measurements reported from stations, OGC geographical web services and geocatalogs (Figure 2.10) [HIN 09].

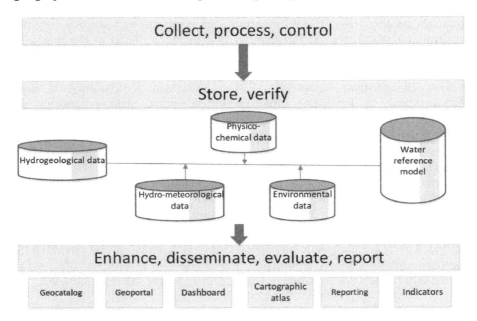

Figure 2.10. *General framework for water information management by the SANDRE initiative*

The WIS in France is based on the National Water Data Scheme (NWDS) [EAU 10]. It defines: (1) the objectives, scope and governance methods of the WIS; (2) the collection, retention and dissemination systems of data and indicators; (3) the implementation methods of these systems; (4) the procedures for developing methodologies and the data and services repository that these systems must respect to ensure their interoperability; (5) the data exchange methods with other information systems which are not fully included in the perimeter of the WIS. Thanks to this legal, organizational and technical basis, the WIS plays an important role in the dissemination and reuse of water data, in particular for decision-makers and managers in the sector. Also, the WIS in France feeds the European System WISE. In France, the WIS operates with the knowledge of certain limits and constraints in relation to the density of the system and the complexity of the storage and aggregation of data from a number of operators in the water sector in France. These limits were identified and communicated during the Eaufrance workshop which was organized in 2011.

2.4.3. WIS in Spain

In Spain, the Public Information Access Right Act was adopted in November 2013. Following dramatic events due to flooding, all drainage basins have been equipped with telemetry stations and control centers for real-time monitoring. In 2000 the portal for access to water information (HISPAGUA: http://hispagua.cedex.es/) was launched at the initiative of the European Water Directorate. This was accomplished in partnership with scientific research centers and bodies, with the support of the European Commission and the EMWIS (www.semide.net). It was the first portal on water information in the Mediterranean region. HISPAGUA is structured in five sections: (1) Institutional water information, (2) Documentations, (3) Research and development in the water field, (4) Training, and (5) Water data bank.

HISAQUA is part of the SIA in Spain (Systema Integrado de Informacion del Agua), inaugurated in 2008 by the Ministry of the Environment, as the official water information space. SIA is composed of three components: (1) an Intranet, to which access is reserved for specialists and managers of the sector; (2) a geoportal accessible to the general public, including several features like thematic map consultation and geospatial data catalogs with the possibility to download water data in different formats; and (3) dynamic electronic books that the user can build from the automatically updated data and according to the available themes (climate change, precipitation, water qualities, etc.).

The purpose of SIA is to guarantee a greater transparency to the general public and also to international partners, in particular the European Union in the framework of the WISE system. SIA puts at the disposal of the general public a geoportal which groups together a set of geospatial data, a digital atlas on water resources and an updated inventory of dams and access to water indicators at national level.

2.4.4. *WIS in Algeria*

The national WIS in Algeria called "La base Eau" (the Water Base) was initiated in 2002. In fact, the historical context of WISs in Algeria has its roots in 1995, during the development of the National Water Strategy (NWS). The NWS has been at the origin of the creation of basin agencies since 1996. It was by registering the establishment and updating of databases and geographic information tools in the list of missions of basin agencies that the information water system could be developed [MOZ 13]. Then a new step was taken for WISs in Algeria, with the adoption of the 2005 Water Act through Article 66, which introduced the integrated water information management system and identified the Ministry of Water Resources (MWR) as the National Focal Point in charge of the WIS. Then, in 2008 Decree 08-326 set the organizational and functioning methods of the integrated water information management system. In 2011, a ministerial decree was adopted to define access rights to water data.

From a technical point of view, according to the progress report of the Algerian WIS, established by EMWIS in 2005 [SEM 05], the information system is composed of a data center at the Ministry of Water Resources (MWR) level, which groups together in a central database under an SQL server all the data collected from basin agencies. The WIS also integrates GIS tools with a web interface. All these elements are associated with this system to facilitate the exchange and consultation of data and thematic maps by all concerned operators.

Each hydraulic basin agency (HBA) has its own information system, including the EMWIS focal point and the Constantine HBA. Each HBA is required to collect water data at basin level. These data collected by agencies concern the data specific to the agency and also those of regional operators concerned with water information. All data thus collected is automatically replicated to "the Water Base" at the central level, accessible via Intranet to all operators at national level (agencies, offices, ministries, etc.). It was in 2010 that this system was put into action [SEM 05, MIN 12]. To facilitate the exchange of data between partners, an interinstitutional protocol has been established as a formal request to receive water data [TIR 12]. However, the slowness of this protocol and the absence so far of access for the general public are an obstacle to the development of a WIS in Algeria and a failure of the predefined vision in collaboration with international partners.

2.4.5. *WIS in Tunisia*

Tunisia, like most North African countries, is under significant pressure in regards to its surface and groundwater resources. This water scarcity situation was the main motivation for the implementation of a series of actions and reforms, beginning with the 1975 Water Code (Act No. 16-75), which constitutes the legal framework for water resources management. Then, in 1998, the water sector adopted its first long-term strategy (Water Sector Strategy in Tunisia in the Long Term 2030) [MEM 13], which identified the necessary financial resources, and which was subsequently named "Water Sector Investment Programs (WSIP)" [MIN 08]. As regards the right of access to the administrative document, Law 2011-41 [REP 11] defines the principles and rules governing access to documents of public bodies.

The national WIS in Tunisia was created following the study carried out in the framework of the first phase (2001–2007) of the WSIP I program. A national WIS is considered as a unifying tool for the information systems of all the operators of the water sector, due to the multitude of databases and tools (mapping, GIS, etc.) implemented by each operator separately. The purpose of the national WIS is to support the process of preventive decision-making and facilitate sharing and communication of information between different national and international partners and operators.

According to the national WIS terms of reference, it should consist of the following three subsystems: the water resources management system, SYGREAU; the water pollution control system, COPEAU/SPORE; and the irrigated perimeter soil quality monitoring system, SISOL. Other modules and systems will have to be integrated into the WIS system, to constitute the final vision of the national WIS project. To make this vision a reality, an agreement was adopted in 2010 for the exchange of data between five ministries concerned with water information. Other actions are underway to finalize the national WIS in Tunisia (commitment of different operators, definition of a reference dataset, constitution of a national committee and working groups, etc.).

2.4.6. *Others*

2.4.6.1. *North America*

The USGS (United States Geological Survey) is the geological commission of the United States, which developed the national water information system referred to as USGS Water Data for the Nation. The USGS national WIS collects and stores water data from 1.5 million sites. This data is automatically saved, then transmitted to the GEOSS satellite at quite regular intervals (approximately one hour), and then sent back to the storage and scientific processing centers in order to finally upload the result online (http://waterdata.usgs.gov/nwis).

2.4.6.2. *Australia*

AWRIS (Australian Water Resources Information System) is the WIS in Australia launched in 2003 in the framework of the Water Act. In 2007 the Bureau of Meteorology of Australia put in place its network of collaborators to ensure the collection, storage, analysis and monitoring necessary for this resource (http://www.bom.gov.au/water/about/wip/awris.shtml), more than 200 agencies and organizations and more than 35,000 sites and stations. The Bureau of Meteorology (BOM) subsequently launched its geoportal to make the data available to the general public, in the framework of the 1982 Public Information Access Right Act, the "Freedom of Information Act" [AUS 82].

2.4.6.3. *Brazil*

The WIS of Brazil, HIDROWEB (Sistema of Informaçoes Hidrologicas), was introduced by the 1997 Water Act and its developments only began in December 2007. The data collected and processed from 17,000 sites and stations are then made available on the HIDRO portal (http://hidroweb.ana.gov.br/). XML is the exchange format adopted between different system components and participants to ensure interoperability.

2.4.6.4. *India*

In 2008, a government agreement led to the creation of the Indian WIS (India-WRIS), with the aim of implementing a Spatial Decision Support Tool (SDSS). The mission of the India-WRIS is defined through the National Water Policy: to provide a single access point to all water resource data (http://www.india-wris.nrsc.gov.in/) in a standardized format, allowing the user to access, view, interpret and analyze data for better management, planning and evaluation.

2.4.6.5. *Africa*

Africa is a continent rich in water resources and with a very young and growing population. But this continent does not currently have an operational WIS that is fully open to the general public, either at the continental or regional levels, except for a few experiments at the national level, as concluded by Prestige Makanga and Dr. Julian Smit in 2010 following the study carried out on continent-wide spatial data infrastructures [MAK 10]. Nevertheless, we find several experiments, initiatives and projects. Among these initiatives is "the African Convention on the Conservation of Nature and Natural Resources", called the "2003 Maputo Convention," established between the member states of the African Union. Article 16 establishes "the right to information on the environment" for the general public and Article 22 specifies that the exchange of information between bodies competent in this field is one of the forms of cooperation between the Member States. In 2007, the African System for Documentation and Information on Water (SADIEau) was launched following the initiative of a set of North–South institutions and partners. SADIEau (www.sadieau.org/) is co-financed by the European Union and the French Government and is led by a consortium of partners including the ANBO (African Network of Basin Organizations) and ODSR (Organization for the Development of the Senegal River). The SADIEau portal was launched in 2009, grouping together knowledge on water on the continent (more than 500 documents, web pages and links) and a newsletter to disseminate news and information. However, this portal does not have georeferenced data or a geoportal.

2.4.6.6. *Canada*

The Groundwater Information Network (GIN) is a Canadian GDI for the management of groundwater resources. It enhances the knowledge of groundwater systems and facilitates management through access to standardized information. GIN uses the GDI's own standards and technologies to offer transparent access to distributed and heterogeneous data. Its sophisticated portal (www.gw-info.net) allows visualization and analysis of data in order to meet scientific, commercial and water management needs. GIN represents one of the first GDI in hydrogeology [BRO 16].

2.4.6.7. *Europe*[1]

Developed as part of the implementation of the European Water Framework Directive, eWater represents a GDI for hydrogeologic data management in Europe. Its main objective is to increase the accessibility and use of spatialized data on the quality, location and use of groundwater. This project facilitates the exchange of data between national and regional water resource management agencies, as well as the resolution of cross-border water management problems.

In addition, eWater aims to strengthen the partnership between hydrogeologic data providers by better harmonizing metadata on spatialized hydrogeological information and by accessing data and services over the internet.

2.4.6.8. *Austria*[2]

The Water Information System Austria (WISA) is the central platform for data and information on national water management in Austria. It provides public access to data, maps and documents on the implementation of EU water directives, such as the Water Framework, Floods, Groundwater or Urban Waste Water Treatment Directives. WISA is the official platform for the publication of Austrian hydrographic basin and flood risk management plans.

WISA makes water information accessible to interested members of the public, national ministries, regional offices and experts. It supports water management planning and administrative processes at federal and provincial level, as well as compliance with national and international reporting obligations and public participation requirements.

WISA integrates data from various data sources and information systems. It is based largely on already well-established information flows between national and regional institutions. In addition to a central data warehouse for aggregated data, WISA provides access to sectoral databases containing disaggregated data. From a technological point of view, the WISA geoportal is in conformity with the OGC and ISO standards.

---

1 This section appeared previously in an article by the author [JAR 15].

2 This section is quoted from the following source: http://www.umweltbundesamt.at/en/ services/services_resources/services_water/wisa_en/.

2.5. Water data standards

The efficient and rational management of water resources is largely dependent on the availability of large quantities of very good quality data. Putting all the data in a coherent and logical structure supported by an object-relational database helps to ensure validity and availability. This makes it possible to provide a powerful tool for water studies.

From an organizational point of view, access to and exchange of water information in the framework of the implementation of a GDI depend mainly on data models, which can vary from one organization to another [WOJ 13]. In order to make information available and accessible, it is essential to consolidate these models into a single model.

The objective of this section is to present an existing water model package which contributes significantly to the standardization of water information as well as to promoting the development of a geospatial water data infrastructure [BRO 11]. These water data models, which are presented here, must ensure, at national level, the homogeneity of water information. They are described by a series of Unified Modeling Language (UML) diagrams using object-oriented modeling and following the ISO/TC211 recommendations and the Open Geospatial Consortium (OGC).

2.5.1. *Water data acquisition standards*

Water data acquisition standards include (Table 2.2) field observation methods and laboratory analysis methods for water samples. Field observation procedures include those that allow the measurement of precipitation, river flows, groundwater levels and various water quality parameters. In the case of water quality data, there are also standard methods for laboratory sample analysis. A selective sample of some of the most important field observation and laboratory analysis standards is listed [WMO 17].

Documents	Authors	Date of publication
Standard Methods for the Examination of Water and Wastewater	American Public Health Association	2012

Guide to Meteorological Instruments and Methods of Observation.	World Meteorological Organization (WMO)	2008
Guidelines and standard procedures for continuous water-quality monitors – Station operation, record computation, and data reporting.	U.S. Geological Survey Techniques and Methods	2006
Groundwater technical procedures of the U.S. Geological Survey	U.S. Geological Survey Techniques and Methods	2001
Australian Guidelines for Water Quality Monitoring and Reporting	Australian and New Zealand Environment and Conservation Council; Agriculture and Resource Management Council of Australia and New Zealand	2000

Table 2.2. *Main standards for water data acquisition*

2.5.2. *International models and standards for water data exchange*

In the hydro-informatics field, object-oriented methodology can be considered as a new solution to reduce the relative complexity of water data modeling. The UML formalism must be used to develop object-oriented conceptual models of water information, and then to describe their structure from different points of view and at different stages of development. Currently, UML notation is used in many different areas ranging from business process description to environmental issues such as hydrology or hydrogeology [WOJ 13].

Conceptual modeling in geomatics does not require all the methodologies and possibilities of UML notation. A narrower geomatic profile can be achieved, including technical specifications accepted by ISO/TC211 and described in ISO 19103 [ISO 15a], with additional information in ISO 19109 [ISO 15b], 19118 [ISO 11b] and 19136 [ISO 07]. Provided that these standards are met, existing research, analysis and visualization tools can be reused. Geographic objects encoded according to ISO/TC211 and OGC are easily exchangeable for different users, regardless of the proprietary or open source software used.

The UML developers wanted to address different architectural complexity levels and different possible application areas. Some of the core benefits of using UML as the standard conceptual schema language for water data modeling are the following:

– use of a common language between computer scientists and professionals who operate in the water sector;

– compliance and perfect accord with ISO/TC 211 standards, as well as the standards published by OGC, which require the use of the UML notation for the development of geomatic applications;

– sharing and exchange of information interoperably between project operators will be possible by using different web services for data search and delivery.

All existing models in terms of exchange and sharing of water data have a different orientation, driven by a particular need for standards in a particular context. This following sections provide a global overview of the standards and references deemed relevant for the implementation of a GDI on water.

2.5.2.1. WaterML

The Consortium of Universities for the Advancement of Hydrologic Science (CUAHSI) has developed the WaterML standard [DEL 14], now in version 2.0, which allows the coding of hydrological observations via their web services, WaterOneFlow. The first development goal of WaterML 2.0 was "to encode the semantics of the discovery and recovery of hydrological observations and to implement both generic and unambiguous hydrological data services for different data providers and the hydrological research community".

WaterML 2.0 is implemented as an XML schema and currently uses OGC and ISO. The semantics used come from the CUAHSI observation data model. Version 2.0 marks a harmonization with different formats of various organizations and countries, including the Australian Water Data Transfer Format, WaterML 1.0 of the United States, XHydro of Germany and existing OGC formats. WaterML 2.0 was adopted as the official standard by CMO in September 2012, approved by the Federal Geographic Data

Committee (FGDC) of the United States, and has been proposed for adoption by the World Meteorological Organization (WMO).

2.5.2.2. *Water data and repository administration service (SANDRE)*

The role of the SANDRE initiative was to develop a common language for interoperability of water data in France. Before the creation of SANDRE in 1993, water data were heterogeneous. Each producer defined their own codification, chose their definitions, applied their nomenclatures and structured their files according to their own standards. Since 2010, France has defined its National Water Data Scheme (NWDS) laying down the objectives, scope and governance methods of water data, based of course on the SANDRE repository. It is a specification that is based on four document types:

– data dictionaries, which make it possible to define the terminology and the data available for a particular topic (groundwater, water quality, etc.). They also include the association between different objects (the "well" object is for example related to the "aquifer" object). They are used in particular to design databases;

– exchange scenarios, which make it possible to describe methods for the exchange and sharing of water data. They are used to describe semantic and attributary data in detail;

– transformation scenarios (specification documents) are the set of rules allowing the detailed description of the transformation methods of a document from one format to another;

– web service scenarios (specification documents) make it possible to describe methods for the exchange of data via a web interface.

All of these documents facilitate the sharing and exchange of water data by specifying the formats and rules to be applied.

2.5.2.3. *Australian Hydrological Geospatial Fabric (Geofabric)*

Geofabric provides a water data repository that works transparently across Australia. Users can discover, visualize and model hydrological characteristics throughout Australia at the appropriate level according to their needs.

It includes six data categories:

– hydrological regions defining river regions across Australia;

– hydrological drainage basin reports, providing additional details for small waterways as well as a simplified waterway network;

– surface basins defining base-level drainage basins for waterway tributaries, wells and coastal drainage areas;

– surface network providing a fully connected and oriented detailed flow network. The user can follow flow paths and link them to surface basins;

– surface mapping makes it possible to visualize surface water characteristics such as dams, canals and bridges;

– groundwater mapping shows groundwater resources and their characteristics, such as aquifer limits, salinity, and rocks and sediments at different levels below the surface.

2.5.2.4. Groundwater Water Markup Langage (GWML)

GWML is an international standard for the online exchange of groundwater data which addresses the data heterogeneity problem [BRO 11, BRO 16, BRO 18]. This heterogeneity makes groundwater data difficult to discover and use because they are structured and fragmented into many silos. Overcoming data heterogeneity requires a common data format; however, until the development of GWML2, there was not an appropriate international standard. GWML2 represents the main hydrogeological entities such as aquifers and water wells, as well as related measurements and groundwater flows. The standard was developed and tested by an international consortium of groundwater data providers from North America, Europe and Australia; it facilitates many data exchange formats, the representation of information and the development of web portals and online tools.

2.5.2.5. The Arc Hydro model

Arc Hydro is a data model developed by the University of Texas to standardize the representation of spatial and temporal water information. It is primarily used to store, document and analyze spatial and temporal datasets commonly used in the field of water resources management.

Standardization is important because it allows the sharing of data and applications between several water agencies, and provides a framework for integration with hydrological calculation programs of various types. Various software tools have been developed to facilitate database populating with information on drainage basins, hydraulic structures, measuring stations, and properties of the surface of the earth. Geographical features can be linked to time series data in the database, thus providing the basis for simulation modeling for urban and hydrological drainage basins as well as groundwater.

Arc Hydro is structured in three main components:

– a standardized format for the storage of geographical and chronological hydrological data;

– logical data relationships between geographical entities (or "objects");

– a set of tools to create, manipulate and visualize hydrological data.

2.5.2.6. *Water Data Transfer Format (WDTF)*

WDTF is currently being developed by the Australian Bureau of Meteorology and CSIRO as part of the alliance for water information research and development. It is part of the AWRIS software of the Bureau of Meteorology. The scope of the format is to enable information coding which must be provided to the Bureau by water agencies or organizations that carry out hydrological measurements. The standard addressed not only observational data, but also descriptions of characteristics (storage, waterways), transactional information (synchronization with a data warehouse) and water quality samples. The new version 1.2 also includes groundwater observations.

This format uses the OGC specification, through a simple GML profile according to the standard ISO 19125 which limits certain aspects such as available geometries and type complexity. It also uses the GML standards with respect to space objects modeling.

2.5.2.7. *XHydro*

XHydro is a water exchange format which has been developed by the Federal Ministry of Transport and Urban Affairs in Germany with, of course, the involvement of German states and industrial partners. Its purpose is to standardize time series data transmissions between sensors, data measurements, the central database and the long-term data archives.

XHydro is more than an XML-based exchange format for water data and XML schemas the generic time data coding; it also has an extension that is specifically designed for discharge and water level data. The development process has been largely influenced by the development of a technical architecture in order to harmonize water data.

2.5.2.8. *The DelftFEWS published interface*

The DelftFEWS model is used by operational forecasting agencies around the world to perform hydrological forecasts, such as river flow (floods and water supply), groundwater levels, water quality and harmful algae proliferation. To accommodate this range of areas, the DelftFEWS published interface has been defined as a standard for exchanging water data (model parameters, simulation model results). The standard defines time series in XML, time series structured in NetCDF format, metadata on model states and model datasets in XML, while states and model datasets are exchanged in native format.

2.5.2.9. *OGC information technology standards for sustainable development*

This document includes overviews of existing standards, current work on new standards, technological trends affecting environmental standards, and the description of some of the standards that are necessary but do not yet exist. The authors argue in favor of the development and use of domain-specific but technically interdependent IT standards for data communication and integration within and between areas that focus on the environment. These areas include earth sciences as well as environmental response and management areas such as emission monitoring, compensation, trade, taxation and regulation, physical infrastructure monitoring; public health; integrated energy monitoring/reduction, intervention in case of a disaster; etc.

2.6. Conclusion

Water resource integrated management requires consultation and better communication between different operators of the sector. For this reason, geospatial data infrastructures offer the opportunity to produce and federate water data gathered by different operators, for a better use in particular through the contribution of geographic web services and OGC standards which facilitate data exchange in different formats. The importance of GDIs has led most of the countries to integrate this system in their development approach. In this context, GDIs play a major role in scientific research, decision support for water resources planning and management, and the participation of different operators in the decision-making processes. The existence of an infrastructure offering services of discovery and sharing of these georeferenced data becomes an imperative, so that all the operators concerned can test and validate their hypotheses and communicate.

Water data use is greatly improved through the adoption of water data standards and quality management processes. Water data standards include (1) standards that guide field observation methods and water sample laboratory analysis, and (2) standards of water data storage, exchange and transmission. There are many different standards, although it is recommended to adopt international and national standards. The most widely used water data transfer standard is WaterML 2.0, which is published by OGC. This standard is relatively new and continues to develop.

Case Studies

3.1. Cataloging data on groundwater resources

3.1.1. *Introduction*

Water resources is an area in which geospatial information can have a positive impact on improving the decision-making process. In reality, water is part of a complex system whose proper management depends on the means of knowledge and access to information [IST 03]. Managers operating in the water sector must be able to obtain reliable, up-to-date and relevant information, when they need it and in a manner that suits them [DON 09].

In Morocco, over the last two decades, significant progress has been made in the development of a set of water resources information subsystems affecting telemetry programs, water quality monitoring, flood announcement management and other areas.

The report by the national water information system in Morocco (national WIS) shows that it is fragmented into a subsystem oriented mainly towards statistical activities. There is a multiplicity of information systems, most of them unrelated to the others. Data transmission is ascending, heavy, without purpose and consumes a lot of expensive resources. At the central level, it accumulates a disproportionate amount of information which is simply archived and remains inaccessible to a large number of participants in the water sector.

Furthermore, the development of each subsystem took place in an isolated setting without taking into consideration the norms and standards, or the problem of the optimization of the energies deployed for information collection and processing.

Water information subsystems are often fragmented, vertical, not participatory, non-transparent and generally only cover part of the water system, since the private sector is not included.

All of this limits the planning and the use of data for the decision-making process, which is different according to time (short, medium or long term), scale (national, regional, communal), structure, level (central or peripheral, administrative, political) and different operators (decision-makers, planners, professionals, researchers).

To be able to effectively use water resources data, one must at any time be up-to-date, compatible with others and accessible to a wide audience. In this sense, GDIs have must ensure a simple, permanent and advantageous access to spatial information.

Keeping this problem in mind, the basic hypothesis of my research is to develop a technical framework for the implementation of a GDI in the prospect of improving access to water resources data in Morocco and to make these data interoperable.

3.1.2. *Report by the national water information system in Morocco (national WIS)*

Groundwater resources management in Morocco is mainly provided by the Ministry of Energy, Mines, Water and Environment (MEMEE)[1] and more specifically the Department of Water which includes the following bodies:

– Directorate General of Hydraulics (DGH) including the Directorate of Research and Water Planning (DRPE)[2] and the Directorate of Hydraulic Facilities;

– Hydraulic Basin Agencies (HBA), public administrative establishments with financial autonomy;

– Directorate of National Meteorology;

1 MEMEE – Ministère de l'énergie, des mines, de l'eau et de l'environnement.
2 DRPE – Direction de la recherche et de la planification de l'eau.

– National Office of Electricity and Drinking Water (ONEE)[3], a public autonomous commercial institution.

The information collected by the DGH and HBAs are centralized and stored in a water database management platform in Morocco called Badre21 (Figure 3.1). The Badre21 system contains a large part of quantitative water data. It has been developed by the MEMEE Department of Water and operates in a decentralised manner at the level of the nine basin agencies. Badre21 manages the properties of water points and measuring points. Some of the information is published on the Internet, including the situation of storage dams and the hydrological situation at the level of hydrometric stations [MIN 12].

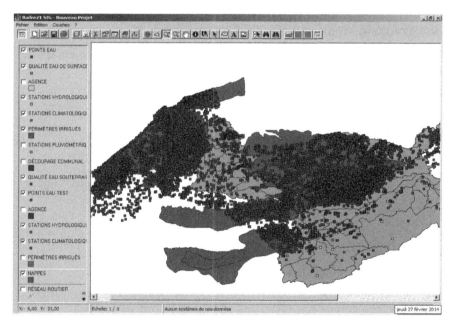

Figure 3.1. *Graphical interface of the Badre21 application. For a color version of this figure, see www.iste.co.uk/jarar/spatial.zip*

On the other hand, specific applications have been developed independently by the HBAs and some divisions of the Department of Water to meet their specific needs, in particular:

---

3 ONEE – Office national de l'électricité et de l'eau potable.

– flood announcement management (BAC21);

– geographic information system (GIS) (Sigaud) of water points (more than 110,000: boreholes, wells, dams, etc.);

– GIS of pollution sources of water resources;

– GIS of water resources quality;

– telemetering system of hydrogeologic data with centralization (DEMAsole software) for subsequent integration into Badre21;

– dynamic geospatial website at the level of Souss-Massa-Draa HBA.

The information system on water resources suffers from several problems, which today makes the sharing and exchange of groundwater resources data in Morocco difficult, if not impossible, because:

– standards are not supported by the implemented software components (for example, Badre21);

– data is not updated regularly;

– data exchange is not based on well-defined protocols;

– there exists a use of different platforms for the exchange of data between several branches of the same ministry;

– there is the lack of a metadata cataloging system for effective openness between partners.

To resolve these different problems, we propose a technical framework for the sharing and management of geospatial data based on GDI concepts.

3.1.3. *The operators*

Operators can be grouped into the following broad categories: data producers, software developers, intermediary, university and users.

Data producers: develop and produce data in different formats (maps, digital elevation model (DEM), images, orthophotos, etc.) and disseminate them through IDS services.

Software developers: develop applications that allow the use of GDI services or they develop geoportals allowing data consultation. These are generally private companies or universities.

Intermediaries (also called brokers): adapt and integrate existing components and solutions to provide a complete and scalable system to non-expert users and organizations. They are normally private companies.

Universities: develop algorithms, methods, programs and solutions that do not exist on the market so that technology can progress and evolve.

Users: use the services provided by a GDI to solve their problems. They ask for information. They may be individual citizens, public bodies, private companies, universities, associations or any social agents. The user is the most important operator of a GDI.

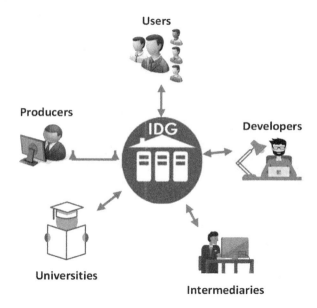

Figure 3.2. *The principal operators of a GDI*

3.1.4. *Material and method*

Regarding this project, the technical choice is oriented towards the "deegree" solution as an online cataloging and publication tool for water resources data. This choice is justified for the following reasons:

– deegree is a turnkey GDI, grouping together a set of modules allowing both to have a geocatalog and a geoportal on the same platform, in addition to a map server of the same family (deegree-WMS and deegree-WFS), with the possibility of the integration of another kind of module (for example, deegree-SOS for real-time data feedback);

– deegree is a GNU LGPL licensed solution, using Apache-Tomcate as a JSP servlet container. This further facilitates the deployment stages and thus allows one to have a prototype whose architecture is service-oriented;

– deegree implements a large number of ISO and OGC standards. In addition, deegree conforms with the INSPIRE directive, which is an important case study for the reuse of this directive in the particular context of Morocco and, at a larger scale, at the Maghreb level;

– deegree has a large user community and represents a major advantage given that it is the outcome of research conducted at the University of Bonn (Germany), which is very adapted to the current situation;

– Finally, this prototype is essentially based on a service-oriented architecture, in particular on the web services concept [POR 08], to guarantee interoperability and ensure the online dissemination of hydraulic and hydrogeological data. The user specifies the search criteria for data. These criteria are transmitted to the JSP servlet server. The result is returned to the user in a user-friendly interface with several XML/PDF extraction features, with the possibility of downloading or consulting online geospatial data according to their formats.

3.1.5. *Architecture of a GDI on groundwater*

Cuthbert [CUT 98] proposes a geospatial data publication model consisting of four processing units and four representation components (Figure 3.3):

– the data selection process is managed by the OGC/ISO "Filter Enconding" specification;

– access to vector data is achieved via the SQL (Structured Query Language), which is offered by the implementation of the "Simple Feature Specification", or with its Web Feature Service (WFS) equivalent;

– the representation of "Features" entities and "Feature Collections" is carried out by GML (Geography Markup Language) standards;

– the "Display" view generator applies SLD (style layer descriptor) style rules to geographic entities and produces graphical representations in image format (PNG, TIFF, JPEG, etc.);

– the implementation of a Web Map Service (WMS) makes it possible to define an interface to access the map which represents the result of two processes: the "Display" view and "Render" compression;

– the obtained result is delivered to the user via a web client.

Based on the Cuthbert model, an interoperable platform based on an n-tier architecture has been developed (Figure 3.3). The overall idea behind this platform is to use the web services concept to provide access to hydrogeologic data.

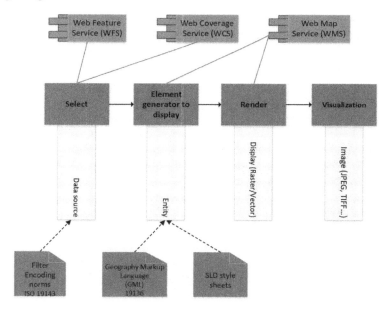

Figure 3.3. *The representation model*

On the server side, the engine of our system is an Apache Tomcat application server, developed in Java. It is in fact a servlet container in which the other components of this tier are deployed. This container also contains a web server that makes it possible to receive queries and send responses to client destinations following several protocols, including HTTP, which is useful to us as it is able to carry out these exchanges over the Internet [DES 10].

The main component of our server is based on the deegree open-source project. It contains the services necessary for a GDI (deegree web service) as well as those for the components of a portal. It offers a mechanism to manage security and access control problems (deegree iGeoSecurity) and storage. It is based on OGC and ISO 19115/19119 standards. It has been used as the main support for the implementation of different web services (Figure 3.4):

– Web Map Service (WMS): the WMS specification describes an interface on which georeferenced maps can be made available [ANN 05]. Data is viewed in image form (gif, png, jpeg, svg). The service consists of three operations to send queries to the server and obtain information: GetCapabilities, GetMap and GetFeatureInfo;

– Web Coverage Service (WCS): proposes a standard for the behavior of a web service for the production of georeferenced matrix data. It describes the structure of the query to be sent to the server and the manner in which the server returns the response to produce maps. WCS servers support the following operations: GetCapabilities, DescribeCoverage, GetCoverage;

– Web Feature Service (WFS): offers direct access to geographic information. The OGC GML data exchange mechanism is used, as the basis for the WFS specifications.

We can differentiate WFS according to two types:

– Basic WFS: a basic WFS contains the GetCapabilities, DescribeFeatureType and GetFeature operations;

– Transaction WFS: supports all basic WFS operations and in addition Transaction and LockFeature operations (optional).

These services are represented by deployable servlets in the Apache Tomcat server. They are naturally linked to other specifications:

– Simple Features (allowing customers to access geographic objects through adapted queries);

– Coordinate Transformation Services (allowing an overlay of information from different coordinate systems);

– Geography Markup Language (allowing a transmission of geometric and semantic information to the client, consequently extending the only visualization of images to the visualization of geographic objects).

One of the key points of the application lies in the connections between web services (WFS, WMS, and CSW) and database management systems (DBMS). In fact, the interfaces proposed to the user must allow a quick and easy selection and consultation of DBMSs where data is stored and must therefore directly exploit the possibilities offered by the SQL. This can be achieved thanks to JDBC (Java Database Connectivity) technology, which is a set of classes allowing the development of applications that are able to connect to database servers (in our case, PostgreSQL). At the level of this project, a connection of our application with PostgreSQL has been established via JDBC technology. This connection allows a dialog between the database and the application. The user interacts with the WFS of the application and, through this, sends queries to the data in the database. These queries are centralized, and then transformed into SQL statements through JDBC technology, which makes the connection with the DBMS and transmits the statements. The results then follow the reverse path until they reach the user. The system uses the OpenStreetMap API in order to have topographic bases for the map display module.

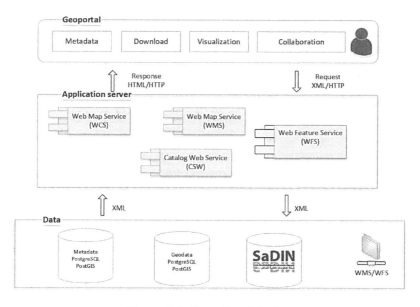

Figure 3.4. *Overall architecture*

3.1.6. *Geoportal*

The implementation of a GDI generally uses a geoportal which is used as a single access point to the infrastructure and which provides discovery, visualization and data access features [FOR 13]. In the framework of this work, a geoportal has been created to be used as an input interface. This geoportal has the objective to provide the largest possible input point to search for the main geographic and alphanumeric data on groundwater resources.

The cartographic space allows us to visualize geographic data (Figure 3.5): we can thus select an area of interest, a scale and the presented themes that a user must present in a map (such as drainage sub-basins, water points, fields, etc.).

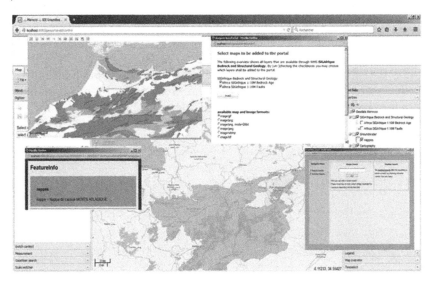

Figure 3.5. *Graphical interface of the Moroccan GDI. For a color version of this figure, see www.iste.co.uk/jarar/spatial.zip*

In addition to the standard zoom, move or query features, it is possible to search for, select and navigate to geographic entities (boreholes, hydrographic network, drainage basin, etc.).

Documents such as OGC Web Map Context can be created, stored and then used to find a defined state of the web interface.

Each thematic map includes tools of consultation, move, printing, etc., as well as specific geographic data that are displayed according to scale thresholds and with predefined skins (toponymy, color, size, style, etc.).

Based on database management system technologies, the geoportal offers the possibility to access drilling data as descriptive sheets and/or stratigraphic logs. It also allows us to produce interactive maps according to the user's needs.

3.1.7. *Geocatalog*

The GDI also integrates a cataloging of all the georeferenced data via a search engine, allowing the identification and choice of data that the user wishes to view. Metadata is structured in accordance with ISO 19115 and ISO 19139 standards for metadata structuring and coding.

This geocatalog uses CSW queries to query the server and retrieve the searched data, based on the metadata that is already entered via the "Editor module" of the Deegree-CSW open-source application. The consultation module makes it possible to perform a search by keyword only or to include other search criteria such as: geographic region and creation/modification date of the searched data.

In order to obtain a solution that includes both a geoportal and a geocatalog, we have carried out the necessary configuration to activate the "CSW-Client" module integrated in the geoportal. The result is shown in Figure 3.6, which presents an example of a search carried out from the geoportal to query the deegree-CSW server on water data of the Ziz-Rheris region, for example: metadata consultation of a layer published on the WMS server, concerning the dams of this area.

3.1.7.1. *Access mode*

The application operates in two access modes, the first in free access and the second in authenticated mode (for the administrator and the editor).

In free access, the user can navigate on different menus of the home page, especially to access the metadata search module.

Only authenticated users can have access to more features: metadata editing, installation and predefined queries console. These users (cmUser, cmEditor and cmAdmin) are predefined in the deegree configuration files.

The different roles for each user:

– the end user (cmUser): the only person who accesses the search module, for consultation;

– the editor (cmEditor): the person who documents data by filling in metadata editor module files;

– the administrator (cmAdmin): the person who accesses the database installation module and the predefined queries console for application maintenance needs. The administrator does not have authorization to access the metadata edit module, only the consultation.

These different access modes make it possible to define three profiles:

– a general public profile, represented by access in "User" mode, with a read-only right;

– an editor or manager profile of datasets, with read and write rights;

– a technical or administrator profile of the application and its database.

What may be interesting in the context of this work is:

– the presence of a profile dedicated to the "general public" in free access and which thus facilitates accessibility to water resources data via an identified and unique information point;

– in addition, the existence of an "Editor" profile of the set of data and metadata via the editor module detailed below, with obligatory authentication. This profile is interesting given the multiple operators who operate in the water sector, for example in the case of Morocco, which requires a more secure access mode that is adapted to each operator's requirements.

3.1.7.2. *Catalog manager*

The catalog manager includes two modules: metadata search module and identified data location (Figure 3.6), through a search engine based on different forms of queries and criteria:

– general query, whose search criterion is a keyword only;

– query by geographic area, to search for the dataset according to a spatial repository;

– query by date, to apply a limit to a defined temporal space.

These three query modes can be combined for more precision and much more relevant results. The objective is to offer the user the tools to target their search.

The editor module: the module that makes it possible to input metadata of associated datasets.

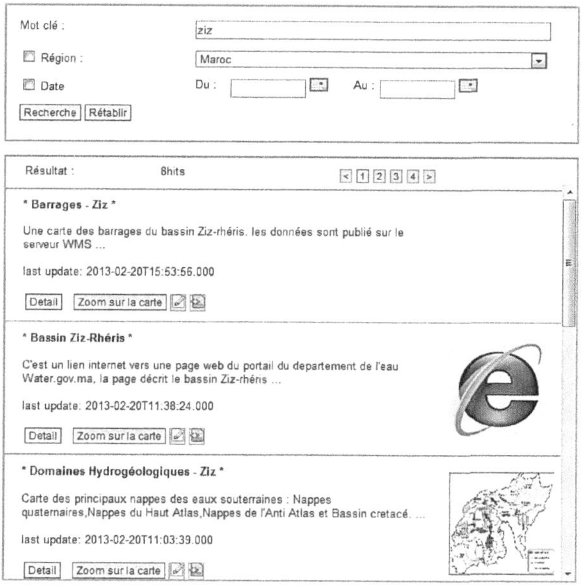

Figure 3.6. *The results of the "Ziz" search in the geocatalog*

The editor module (Figure 3.7) is composed of several series of interfaces making it possible to enrich metadata in accordance with the ISO 19115 standard. Among the interfaces that make up this module are:

– the interface of general information, to fill in the title, a summary and the provider contacts;

– an interface to fill in the temporal information relating to the data (creation, publication and update dates);

– an interface to fill in the spatial information: the region and the coordinate system;

– an interface to fill in the data supplier information; and

– an interface to specify a link to directly access the data or a service (the case of the WMS and WFS layers).

Figure 3.7. *Metadata editor module*

For more details on the selected file, simply click on the detail button to view the details on the metadata of the selected data (Figure 3.8).

Figure 3.8. *Detail of the search result, with a link to the WMS web service to consult the layer published on the deegree map server*

3.1.8. *Conclusion*

The combination of a geocatalog with a geoportal of the deegree family thus makes it possible to enhance data heritage at a lower cost and to direct the public investment effort on more important activities such as the training and upgrading of human resources that will have the burden of enriching the data and their metadata at tool level, which would help to make water resources data more accessible and thus assist decision-makers, as it was indicated at the time of the establishment of the report of the Moroccan Water IS.

The developed platform is the base of an ambitious project to design a geospatial data infrastructure (GDI) in Morocco for groundwater management.

This application is mainly structured around international standards (OGC and ISO). The data come from different databases (PostgreSQL, Badre21, etc.). The application is connected to these data sources either by JDBC technology or via WMS/WFS. J2E technology has been used for the development of the main components of our application. All document formats used are XML applications conforming with OGC standards (GetCapabilities, GetMap, etc.) as well as W3C standards.

Several business-oriented modules have been developed for the decision-making process:

– a spatial data catalog that uses CSW queries to query the server and retrieve the searched data, based on the metadata that is already entered via the "Editor module" of the deegree-CSW open-source application;

– visualization of groundwater resources;

– etc.

This platform is seen both as a tool for sharing and pooling updated data from different data sources and also as a tool for managing water resources on the web providing the indicators for decision-making.

3.2. Geosensors Sensor Observation Service (SOS) for sustainable water resources management

3.2.1. *Introduction*

The impact of human activities on natural resources is increasingly important and directly influences the availability of these resources on a planetary scale [JON 02]. Water is one of the more vulnerable natural resources to this degree, given that the availability of this resource is decisive for human survival and the continuity of human activities. Hence the need to implement a strategy for the sustainable management of this resource.

Several initiatives, meetings and publications have addressed this problem. We list the Dublin Water Conference in 1992, various world water forums, the EUWI European Initiative and various reports developed by UNESCO and UN working groups – including the report of the UN-Water Task Force group, which proposes a dashboard with a quantitative and qualitative indicator suite. This overview can be used to better guide public policies for water resources management.

In Morocco, the hydrological context has for years been considered rather positive, but this favorable situation is no longer the same since the drought years Morocco experienced (between the 1980s and 1990s). In addition to these dramatic years, the impact of the urbanization rate is added, which reached 51.4% in 1994. Another factor influences this situation: the continued population growth, which was estimated in 2004 at more than 29 million inhabitants according to the PHC. In 2014, it exceeded the threshold of 33 million inhabitants.

Faced with these challenges, Morocco prepared in 2009 its Water Strategy. The purpose of this strategy is to ensure an integrated and efficient management of this vital resource. And among the six components which constitute it, the National Water Information System component is placed as a support process of all the action plans envisaged for the development of this sector. Within this component, the sensor network modernization is one of the essential stages planned for the implementation of a National Water IS. In order to illustrate the importance of this stage, a budget estimated at five billion dirhams (~500 million euros) has been assigned to the measuring network modernization work (6% of the cost, estimated at 82 billion dirhams for the water strategy implementation).

To operationalize this strategy, the AGIRE (Water Resources Integrated Management Support)[4] program is one of the programs developed within the DRPE, within the framework of the German–Moroccan partnership. This program identified, during the 2012 SGI-Rabat workshop, the measuring network modernization stages: first it would be necessary to generalize the acquisition and use of probes even with a manual data input of the piezometric level, temperature etc., to proceed to the stage of input automation and finally to equip these automatically recording probes with measurement teletransmission means. On the occasion of this workshop, an inventory was created [ALA 12] with the following observation: among the 128 automatically recording probes on the eight HBAs that exist in Morocco, 59 (46%) are equipped with teletransmission means.

These automatically recording probes with telemetry equipment make it possible to provide real-time information to better control and monitor the level of shallow and deep groundwater fields. The use of the DEMASole software adapted to the hardware equipment for data transfer at the central level (DRPE) has been identified as one of the actions to be undertaken for measuring network modernization, by the report on the inventory of the Moroccan WIS [MIN 12]. DEMASole exports the reported data (piezometric level, temperature, etc.) in text format. These data are then manually used to extract graphs and dashboards with the appropriate indicators.

However, in view of the National Water Strategy (NWS), one of the objectives is to allow a wide access of water data to the general public. Moreover, all international partners require access to these data and among these partners, we will find the funders that support Morocco in its efforts regarding the NWS implementation (European Union, World Bank, etc.). As a consequence of this situation, there is an obligation for Morocco to have the necessary tools for regular data exchange according to international standards.

The purpose of this project is to propose an open-source solution that conforms with international standards for the exchange and exploitation of geospatial data, reported in real time by a sensor network. This solution will be able to complement the current system and thus make it possible to improve it at a lower cost.

---

4 AGIRE – Appui à la gestion intégrée des ressources en eau.

In the following sections, we will present an overview of the experiences at the international scale of Water ISs, which are compliant with OGC standards, and incorporate a component for real-time measurement feedback from a sensor network. Next, we will present a report of the currently available free and open-source solutions with a comparative performance and an overview of scientific research using these tools. At the end, we will present our solution, its architecture and its operation, and then we will discuss the results obtained from the exploitation of this solution in a particular region of Morocco: Ziz-Rheris.

3.2.2. *Material and method*

In this section, we will address the methodology followed for the implementation of an open-source solution, by implementing the SOS standard for our study region: the 52° North SOS [SLI 05; KRA 05].

3.2.2.1. *Study area: "the Tafilalet Plain"*

The data (piezometric level and conductivity measurements), used to implement our solution implementing the SOS standard, concern one of the most important regions of Eastern Morocco: Tafilalet. The description of this plain indicates that it is particularly morphological and includes four major drainage basins: Ziz, Guir, Rheris and Maider, oriented in the North-South direction.

The choice of this area, belonging to the Guir-Ghris-Ziz hydraulic basin agency, is justified by the following reasons:

– 1st reason: this HBA, until 2014, unlike other HBAs of Morocco, still did not have its own website. Hence the idea to propose a free and open-source solution, allowing data visualization on a web interface, accessible to the general public[5].

– 2nd reason: the importance and role of groundwater for the socioeconomic development of this region [MEM 13]. Hence the input that a tool can allow the recovery and real-time monitoring of piezometric level measurements, with a view of the history of these measurements.

---

5 List of HBA websites: HBA Sebou – http://www.abhsebou.ma; HBA Tensift – http://www.eau-tensift.net; HBA Oum Erbia – http://www.abhoer.ma; HBA Souss Massa – http://www.abhsm.ma; HBA Moulouya – http://www.abhm.ma; HBA Bouregreg – http://www.abhbc.com.

– 3rd reason: the worrying assessment on groundwater quality, published in the report entitled "L'état de la qualité des ressources en eau au Maroc de 2002 à 2008"[6] [MEM 13], which is available from the DRPE, informs us that the average overall quality is negatively evolving, which is an important motivation for the real-time monitoring of quality indicators (for example, conductivity) for this region which has more than 2 million inhabitants (which is more than 7% of the population of Morocco) according to the last general census carried out in 2004 [HCP 04].

– 4th reason: this area is at the heart of several projects carried out in partnership with international organizations (GIZ, JICA, FAO, UN, etc.) which requires the use of recognized standards and norms for the exchange of geospatial data.

Following these elements, this region can be considered as a pilot case for the implementation of free and open-source "web sensing" solutions, which can complement the official existing system.

The following section will provide more details on the overall architecture from a technical point of view, with an overview of the implementation stages of the selected solution.

3.2.2.2. Overall architecture

On the server side, the 52° North SOS service represents the central element. It runs on a Java framework called "OGC web service access framework (OX-framework)" [BRÖ 06; BRÖ 09], which allows the integration of data from different sources and with different formats, while ensuring accessibility to the geospatial web service. The goal sought by the development of this framework is to be able to build flexible architectures. The deployment of this solution is possible on a JSP servlet container, such as the Apache-Tomcat application server.

Before the configuration phase of the SOS server, a first stage is the preparation of the PostGIS database, according to the diagram provided for 52° North and which groups together all the tables designed for the storage of the measurements reported by the sensor [KÜN 10].

---

6 In English: "The state of water resources quality in Morocco from 2002 to 2008".

The configuration phase of the 52° North server is the most important step in the target solution implementation process. In fact, the 52° North configuration files contain a large number of parameters (database URL, user, connection number, table names, etc.) to be initialized for the proper functioning of the SOS service.

The data on the phenomena being monitored by the sensor networks can be transmitted directly via communication networks (Wifi, GPRS, etc.) or stored in a file and then imported in the PostGIS database. This is the scenario which was adopted, since 52° North integrates a data import tool, "Import SOS", which identifies the measurements on the monitored phenomenon, the date and time of the measurement, the unit of measurement used, the sensor name and its geographical position. Then, "Import SOS" proceeds to the import of data and feeds the tables created in the preliminary stage.

The data used (piezometric level and conductivity) come from the literature and from the documentary resources which accompany each project or study of a given area. These reports are available in the DRPE.

After the SOS server, the SOS client provides access to the data reported by the sensor via the "Sensor Observation Service" (SOS) web service on a user-friendly web interface. This makes it possible to view these measurements as a function of time in graph form, and to export the results in different formats (PDF, image, etc.).

At the end of these stages, the URL of the SOS client can be linked either to the geoportal and/or documented on a geocatalog, in order to obtain at the end a more global solution which groups together all the elements facilitating the accessibility, visualization and organization of geospatial data. This result is possible thanks to the data stored in the layer attribute tables published on a map server (via the WMS and WFS web services) and also the metadata stored and used by a CSW data cataloging service.

Figure 3.9 summarizes the overall architecture, which groups together the different systems used to build a modular, open-source solution which conforms with OGC standards. There we will find the user interfaces (Geoportal, SOS Client, geocatalog), the WMS, WFS, SOS and CSW servers, with the different associated modules (metadata editor) and also the PostGIS database with the "Import SOS" module.

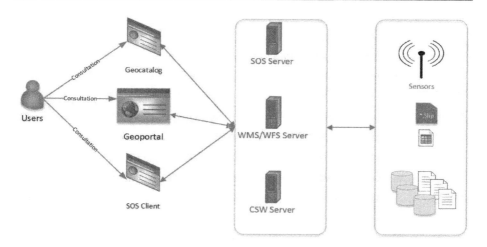

Figure 3.9. *Service-oriented architecture (WMS/WFS, CSW and SOS) of the proposed solution*

The next section presents in detail the results obtained by implementing a set of OGC web services using free and open-source tools for the real-time monitoring of key indicators.

3.2.3. *Results*

In this section we will detail the results obtained after the integration of the data relating to the selected study area, using the 52° North SOS framework.

Two indicators were selected, for the visualization of the results:

– the piezometric level, which provides a quantitative overview of groundwater level, important information to take into account, for example, during the preparation of reports and summaries produced periodically by the organizations operating in this field;

– conductivity, which is one of the qualitative indicators taken into account at the water quality evaluation system level. In fact, the 1275-01 decree of 17 October 2002 sets the quality grid relating to surface water and groundwater, and thus determines the intervals with the associated qualification.

After the implementation of an SOS server and an SOS client interface, the publication of these data on a geoportal of the deegree family consists of editing the attribute data of the measuring station layer, with the QGIS V1.8 office open-source GIS tool, to add the URL of the SOS client. Then, at the geoportal level, it is sufficient to generate a "GetFeature Info" query of the deegree WMS server on the selected station (Figure 3.10).

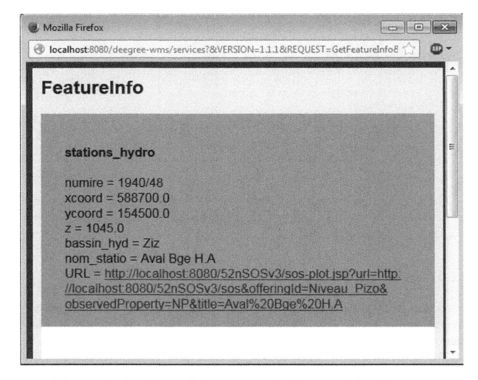

Figure 3.10. *"GetFeatureInfo" query on the "Aval Bge H.A" station, the URL field contains the link to the SOS-Plot interface, to display graphs*

Figure 3.11 represents the history of the piezometric measurements on a graph generated by the Plot SOS library for the "Aval Bge H.A" station of the Ziz-Rheris region, according to the existing data projected over the period of March–April 2014, to simulate the result.

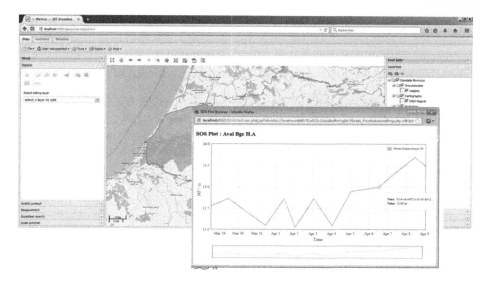

Figure 3.11. *Overview of the piezometric level reported from the "Aval Bbe H.A" station of the Ziz-Rheris region, on the geoportal. For a color version of this figure, see www.iste.co.uk/jarar/spatial.zip*

With these two graphs, the user can read information on the quantitative and qualitative situation of water resources (piezometric level and conductivities), from the measuring station.

To further facilitate the consultation of these data, in the next section we describe the link established between the "SOS Client" interface with the geocatalog, to constitute a complete and homogeneous solution for the same study area.

3.2.4. *Discussion*

In this chapter, we have been able to demonstrate the importance of the use of open-source tools that conform with international standards for real-time feedback on measurements used to better control, monitor and predict groundwater resources.

The proposed solution perfectly integrates into the system implemented by the public administration in charge of water resources management in Morocco. This telemetric system at the end of its operating cycle generates measurements in text format, which can easily feed our proposed solution and thus achieve an automation of all the processing up to uploading of the results on the Internet. Given that the existing solution requires manual processing to highlight graphs and indicators from the data reported by the telemetric system in place using the GSM network, instead of the Internet.

In Morocco, the "IMPETUS Atlas du Maroc" project [SCH 08] is one of the reference research works in the decision support field for sustainable water resources management (http://www.impetus.uni-koeln.de/iida). This project, carried out thanks to the support of international partners of Morocco and mobilizing a large number of research teams, aims to provide decision-makers with the data and means to enable them to make more effective decisions, taking into account a large number of factors (climate, availability of resources, health, society, etc.). One of the IMPETUS components is the use of an open-source solution for water resources data cataloging: the "GeoNetwork" (http://geonetwork-opensource.org). However, IMPETUS does not incorporate a solution for the real-time monitoring of key indicators, which is considered as an added value of the work topics of this research project.

This research work is a continuation of this work (Impetus Project) and proposes a solution incorporating both a geoportal, a geocatalog and the means to monitor in real time the measurements reported by the implemented sensor network. All these solutions are open-source and conform with international standards in the field.

The international situation of Morocco requires the adoption of a policy of openness and exchange of data, which implies the use of international norms and standards. The main partners of Morocco have defined the rules that govern the exchange of information, starting with Article 19 of the Universal Declaration of Human Rights, then the Aarhus European Convention on access to information, and also recently in 2013, the American Open Government Initiative for public data openness (http://opengovernmentdata.org/).

In this international context, Morocco introduced Article 27 in its newly adopted constitution in 2011, which gives the right to public information to any citizen. Also in 2013, Morocco prepared government bill No. 13/31 on the right of access to information which establishes the procedures for requesting information held by public administrations, but does not detail the means that must be put in place to verify first that the said administrations have the information requested by the citizens. In this context, we note the importance of implementing mechanisms to make information available to the general public in real time.

Among the planned future work, in order to continue the current efforts, is the launch of an initiative called the "Water Resources Data Sharing Platform" (PPDRE), which will group together the results of this research work, for the uploading of a portal enabling the general public, and in particular the scientific community, to share water resources data, to enhance their value by allowing the generation of thematic maps on a geoportal, and to facilitate accessibility to these data via a geocatalog, in accordance with OGC standards.

Also, a performance test of this solution coupled with the current telemetric system is among the future prospects for this work. Certainly, we have used a number of data from the available documentary resources, but an operational and performance test of the overall solution (geoportal, geocatalog and SOS client/server) is to be planned with the data reported directly by the automatically recording sensor network deployed at national level.

3.2.5. Conclusion

The main purpose of this project is to propose a low-cost, open-source solution that conforms with international standards and can be integrated into the current system, thus allowing the user to make data available on groundwater fields level and quality measurements.

The use of 52° North, associated with a geoportal and a geocatalog of the deegree family, to build a solution with an SOA, has achieved the overall objective of this work.

The enhancement and the provision of groundwater resources data of a particular region allows a greater reactivity on the part of those in charge and an important implication on the part of the population in projects and decisions affecting the daily life of the inhabitants of this region.

This work is part of the dynamics of developments that take place in Morocco, since the adoption of the Water Strategy in 2009, up to the government bill on the right of access to information prepared in 2013.

3.3. GDI and water data geoprocessing

3.3.1. *Introduction*

The 2012 status report of the WIS in Morocco [MIN 12] shows a diversity of operators in the water and environment field, which inevitably leads to a redundant data production in these sectors and a diversity of their formats and sources. In addition, this generates additional costs as well as a considerable loss of time when searching for reliable data. The main danger is the generation of incomplete and multiplied analyses not allowing one to take full advantage of the data analysis capacities, omitting some aspects for the benefit of others that may be of lower priority. Especially as the report also notes a lack of coordination between operators, which harms and ultimately hinders a proper decision-making process and especially in times of crisis. Also, the cost of the implementation of an IT solution to develop the WIS in Morocco amounts to more than 1.4 million euros and it would be particularly wise to claim the use of free tools and with equivalent performance, even with greater flexibility and a wide choice of software.

There is undeniably an interest to use free and open-source (FOSS) tools in a geospatial data infrastructure (GDI) in a z service-oriented architecture and to implement Open Geospatial Consortium (OGC) and ISO standards. In fact, such an architecture will ensure interoperability between different systems and the same goes for the sharing of different data. In addition, the implementation of a GDI in the water and environment field will avoid the duplication of efforts and expenditure by allowing users to save financial and time resources during data search, acquisition and update [GIU 11b].

In Morocco, several research projects [MOU 16] have demonstrated the importance and interest of implementing these free GDIs in the water field. Their work has made it possible to highlight the feasibility of centralizing and sharing reliable data [MOU 16] thanks to the use of OGC standards, especially the Web Catalog Service (CSW) standard. This is done in order to contribute to better water resources management [JAR 09].

However, these GDIs carried out so far in Morocco had limited analysis capabilities. Now, online geoprocessing transforms data into information essential for the decision-making process. This observation is supported by international research results which demonstrated that during geoprocessing, the limited capacity of desktop computers posed a serious limit to the data analysis range, the quantity and resolution of which continues to grow. This growth is even more marked in the environmental field and therefore must provide a vital capacity to proper decision-making. This, while keeping in mind that the analysis and modeling in the water and environment sector requires the use of a large number of geoprocessing operations and data of different sources and formats.

All the elements provided above lead to the very essence of this work whose primary interest is to achieve a smart GDI using free software which implements OGC standards, especially Web Processing Services (WPS). The GDI will allow one, on the one hand, to achieve the WIS objectives while optimizing financial resources. On the other hand, it will enable online geoprocessing operations and thus extend data analysis capacity. Consequently, through the use of free and effective tools, this work aims to strengthen existing national strategies, especially the NWS which aims, among other things, to modernize information systems in the water sector. Finally, working to achieve the ultimate goal of better integrated and sustainable water and environment management is the context of the research stages of this work.

3.3.2. *Material and method*

Based on the GDI literature, [STE 12] distinguished four components to be addressed during the creation and implementation of an IDS:

– the political component;

– the technical component;

– the commercial component; and

– the social component.

This section will mainly address the technical component.

For the creation of the IDS prototype, the choice of the map server is focused on the free deegree solution with the GNU LGPL license. Among the advantages of deegree is that it provides the main elements for the construction of a free interoperable IDS. In fact, it implements a large number of ISO and OGC standards. Also, this solution provides the user with a geoportal which displays the data served by web services: Web Map Service (WMS), Web Feature Service (WFS), Web Coverage Service

(WCS), etc. Also, deegree implements other OGC standards, including the Web Processing Service (WPS).

deegree certainly offers a WPS server to deploy, publish and standardize the inputs/outputs of processing services of any type of data, including geospatial, on the web. However, in this case of the interpolation of environmental data, the adopted methodology uses the INTAMAP web service [WIL 11] which uses the 52° North WPS server.

The objective of INTAMAP is to automatically interpolate environmental data. It is a free and interoperable web service that implements international standards such as the XML exchange format and the WPS geoprocessing standard. It uses the free and open-source environment of R statistics and geostatistics. The use of R commands via the TCP/IP protocol is achieved through Rserve. INTAMAP implements several geostatistical interpolation methods such as: spatial Copulas, the Projected Spatial Gaussian Process (PSGP) and AUTOMAP, which is an ordinary kriging using the gstat module. INTAMAP also offers the possibility to integrate and develop additional interpolation functions as needed. It is in the framework of the INTAMAP project that the Uncertainty Markup Language (UncertML) interoperable format has been developed. UncertML is a conceptual model and XML coding that was designed to encapsulate the probability distribution of interpolation errors [WIL 11].

Thus, by using an SOA architecture and deegree which implements ISO and OGC standards, this prototype guarantees interoperability and ensures hydraulic and hydrogeological data dissemination. In addition, the WPS component and particularly the use of INTAMAP will make it possible to interpolate environmental data with a simple and installation-free software by the user. This intuitive online geoprocessing integration will help in the decision-making process and proper water resources management.

3.3.3. *Solution architecture*

One of the main objectives of the proposed approach is to demonstrate the feasibility of the implementation, in the Moroccan institutions, of an infrastructure capable of sharing data in an interoperable, simple and free way. Therefore, the proposed architecture (Figure 3.12) takes these conditions into account and proves that with few financial resources, it is possible to share reliable data and perform online geoprocessing with simple mouse clicks, intuitively, without the need for software installation and without advanced knowledge of everything related to GISs.

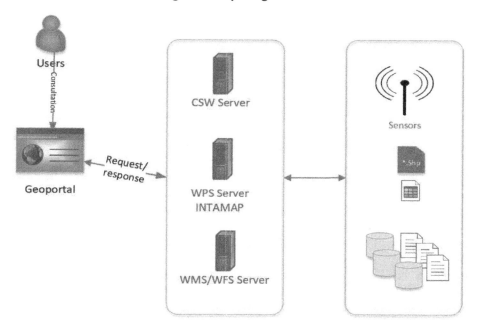

Figure 3.12. *GDI technical architecture integrating an INTAMAP geoprocessing service*

The technical architecture of the solution, as shown in Figure 3.12, is the following three-tier oriented service:

– Tier 1: the DBMS used is PostgreSQL with its PostGIS spatial cartridge. This spatial database contains, in addition to data, all the metadata information entered via the CatalogManager deegree editor.

– Tier 2: the web server used is Apache-Tomcat of the Apache Software Foundation, where the map servers were deployed: deegree-WMS, deegree-WFS and deegree-WCS, in addition to the deegree-CSW geocatalog and the INTAMAP solution with the 52° North WPS server.

– Tier 3: the client interface offers a geoportal for data display and consultation. It allows access to the geocatalog for metadata input and research as well as the possibility to carry out online geoprocessing with simple clicks. In this case, it is the interpolation of a set of one-off measurements through the INTAMAP web service.

3.3.4. *Results and discussions*

The "igeoportal-std" geoportal of deegree version 2.6 is the main interface which allows the consultation of geospatial data published with deegree-WMS, deegree WFS and deegree WCS. It provides the user with the basic features of a geoportal. Figure 3.13 shows the geoportal graphical interface of this prototype, displaying the study area hydrographic network superimposed on the OpenStreetMap mapping background.

Also, this interface offers several complementary modules with the possibility to activate and deactivate them as needed, such as the path calculation module. The deegree-CSW deployment in the web server gives access to the geocatalog manager from the same geoportal. This geocatalog offers the possibility to integrate metadata in the IDS and therefore to know the reliability, sources and production dates and updates of the data. This metadata is in accordance with the ISO 19115 standard. The geocatalog manager offers two modules: the search module for metadata consultation and the editor module for metadata input and edit.

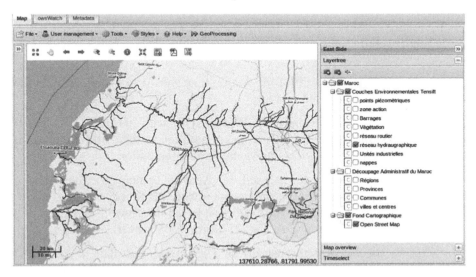

Figure 3.13. *Geoportal graphical interface. For a color version of this figure, see www.iste.co.uk/jarar/spatial.zip*

In addition to the centralization and sharing of reliable data through the previously mentioned elements, the major addition to this GDI is the ability to carry out online geoprocessing and analysis using an OGC standard: the WPS. The one produced in this prototype is the interpolation process carried out with the INTAMAP web service.

In fact, the geoportal offers the possibility to interpolate data with a simple click on the "GeoProcessing" button. This click gives access to the form (Figure 3.14). The first ListBox makes it possible to choose the interpolation method offered by INTAMAP (AUTOMAP, Copula, PSGP, etc.). Then, the user must choose the prediction types to generate. Finally, the user must attach the CSV file (comma-delimited) containing the coordinates of the points and the values to interpolate.

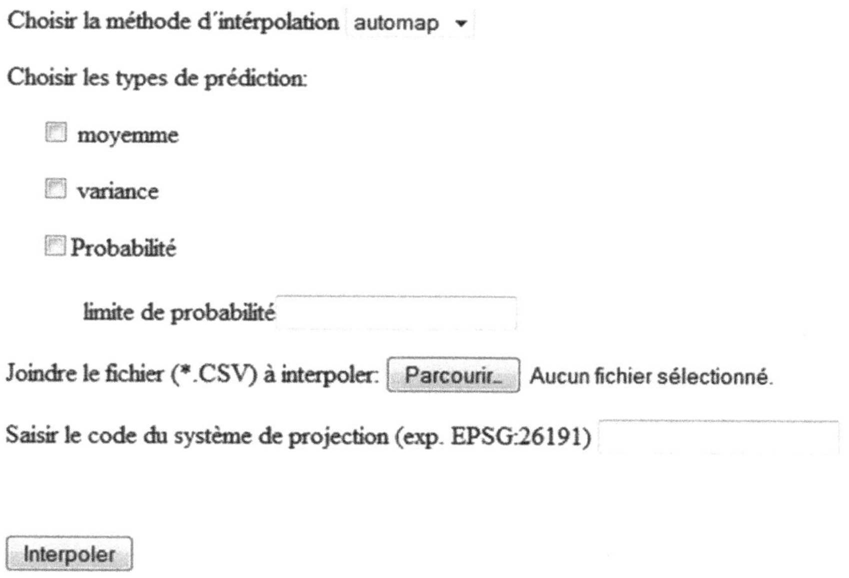

Choisir la méthode d'intérpolation automap ▾

Choisir les types de prédiction:

   ☐ moyemme

   ☐ variance

   ☐ Probabilité

     limite de probabilité

Joindre le fichier (*.CSV) à interpoler: Parcourir... Aucun fichier sélectionné.

Saisir le code du système de projection (exp. EPSG:26191)

Interpoler

Figure 3.14. *Parameter form for the interpolation*

The interoperable INTAMAP web service takes as input the values encoded in the form of the Observations and Measurements (O&M) OGC standard. This prototype supports the conversion internally in the CSV file to the O&M schema and integrates it into an XML file that it creates. This XML file represents the geoprocessing query that the program sends automatically to the WPS server (Figure 3.15).

```
request.xml ×
<?xml version="1.0" encoding="UTF-8"?>
<ns:Execute version="1.0.0" xmlns:ns="http://www.opengis.net/wps/1.0.0">
  <ns1:Identifier xmlns:ns1="http://www.opengis.net/ows/1.1">org.intamap.wps.Interpolate</ns1:Identifier>
  <ns:DataInputs>
    <ns:Input>
      <ns1:Identifier xmlns:ns1="http://www.opengis.net/ows/1.1">ObservationCollection</ns1:Identifier>
      <ns:Data>
        <ns:ComplexData>
          <om:ObservationCollection xmlns:om="http://www.opengis.net/om/1.0" xmlns:gml="http://www.opengis.net/gml" xmlns:xlink="http://
www.w3.org/1999/xlink" xmlns:sa="http://www.opengis.net/sampling/1.0">
            <gml:srsName>26191</gml:srsName>
            <om:member>
              <om:Observation gml:id="1">
                <om:samplingTime/>
                <om:procedure/>
                <om:observedProperty/>
                <om:featureOfInterest>
                  <sa:SamplingPoint>
                    <sa:sampledFeature/>
                    <sa:position>
                      <gml:Point>
                        <gml:pos>183798.504201 119297.498847</gml:pos>
                      </gml:Point>
                    </sa:position>
                  </sa:SamplingPoint>
                </om:featureOfInterest>
                <om:result>2360.0</om:result>
              </om:Observation>
            </om:member>
            <om:member>
              <om:Observation gml:id="2">
                <om:samplingTime/>
```

Figure 3.15. *Snippet from the XML file containing the interpolation query.*
For a color version of this figure, see www.iste.co.uk/jarar/spatial.zip

The result of the interpolation query is returned in the format of an XML file and in a GeoTIFF form for it to be subsequently published via a map server. A snippet of the resulting XML file is represented in Figure 3.16 below.

```
response.xml ×
<?xml version="1.0" encoding="UTF-8"?>
<ns:ExecuteResponse xsi:schemaLocation="http://www.opengis.net/wps/1.0.0 http://geoserver.itc.nl:8080/wps/schemas/wps/1.0.0/
wpsExecute_response.xsd" serviceInstance="http://localhost:8080/wps/WebProcessingService?SERVICE=GetCapabilities&amp;SERVICE=WPS" xml:lang="en-
US" service="WPS" version="1.0.0" xmlns:ns="http://www.opengis.net/wps/1.0.0" xmlns:xsi="http://www.w3.org/2001/XMLSchema-instance">
  <ns:Process ns:processVersion="1">
    <ns1:Identifier xmlns:ns1="http://www.opengis.net/ows/1.1">org.intamap.wps.Interpolate</ns1:Identifier>
    <ows:Title xmlns:wps="http://www.opengis.net/wps/1.0.0" xmlns:ows="http://www.opengis.net/ows/1.1">INTAMAP Automatic Interpolation</
ows:Title>
  </ns:Process>
  <ns:Status creationTime="2015-01-19T02:19:35.344+01:00">
    <ns:ProcessSucceeded>The service succesfully processed the request.</ns:ProcessSucceeded>
  </ns:Status>
  <ns:ProcessOutputs>
    <ns:Output>
      <ns1:Identifier xmlns:ns1="http://www.opengis.net/ows/1.1">PredictedValues</ns1:Identifier>
      <ows:Title xmlns:wps="http://www.opengis.net/wps/1.0.0" xmlns:ows="http://www.opengis.net/ows/1.1">The predicted values</ows:Title>
      <ns:Data>
        <ns:ComplexData>
          <un:StatisticsArray xmlns:un="http://www.uncertml.org">
            <un:elementType>
              <un:Statistic definition="http://dictionary.uncertml.org/statistics/mean"/>
            </un:elementType>
            <un:elementCount>10020</un:elementCount>
            <swe:encoding xmlns:swe="http://www.opengis.net/swe/1.0">
              <swe:TextBlock blockSeparator=" " decimalSeparator="." tokenSeparator=","/>
            </swe:encoding>
            <swe:values xmlns:swe="http://www.opengis.net/swe/1.0">1689.46 1657.58 1622.39 1583.75 1541.59 1495.91 1446.75 1394.23 1338.59
1280.12 1219.27 1156.55 1092.63 1028.26 964.294 901.699 841.503 784.784 732.643 686.167 646.394 614.269 590.607 576.055 571.055 575.825 590.33
```

Figure 3.16. *Snippet from the XML file containing the interpolation query response.*
For a color version of this figure, see www.iste.co.uk/jarar/spatial.zip

As this prototype is intended for decision-makers and the general public, the above XML files remain internal to the system. And since a picture is worth a thousand words, users have access to GeoTIFF directly in the geoportal superimposed to the existing data and the mapping background.

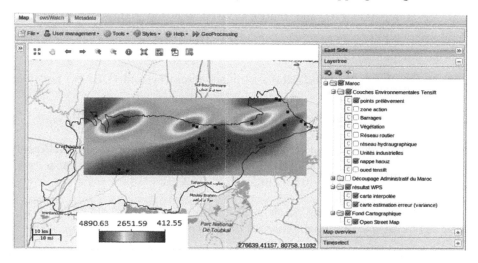

Figure 3.17. *The geoportal showing the interpolation result. For a color version of this figure, see www.iste.co.uk/jarar/spatial.zip*

This generated image will allow the decision-maker to view, quickly and easily, the variation of the electrical conductivity in the HAOUZ fields.

In addition to the interpolation of the values of a a parameter, this system also makes it possible to view the interpolation error estimation image. We note that the error is low at sampling point level. This image does more than identifying the areas where the interpolation error is significant; it shows the user the areas where it is wise to have measured values to generate a better interpolation result.

Certainly, in this example, we have interpolated the piezometric level, but the multidisciplinary INTAMAP web service used makes it possible to interpolate with a simple click all one-off measurements, such as PH, water and air temperature, Cl^-, Na^+ and other parameters.

Thus, without installing a software, without technical prerequisites in GIS, while exceeding the limited capabilities of computers and with simple clicks, the user can query the GDI and execute online geoprocessing queries to make the necessary decisions quickly. The use of the WPS also makes it possible to minimize program maintenance and modification. In fact, [CAS 13] argues that the use of this service makes it possible to:

– be independent of the platform used and the requirements of the computer architecture;

– be independent of incompatibilities between programming languages;

– be independent of the database management system (DBMS) and the version used;

– be interoperable with other existing or future systems.

Thus, the integration of a WPS in the GDI is a stage towards intelligent IDSs allowing an extension of their analysis and decision support capacity – primarily in water and environment areas [GIU 11a].

Several research works, such as Giuliani's in Switzerland [GIU 11a], have led to the use of online geoprocessing using the WPS standard.

In the environmental field, Giuliani [GIU 12] integrated the WPS to calculate the normalized difference vegetation index (NDVI). He used the PyWPS server (http://pywps.wald.intevation.org) which has the advantage of using the GRASS functions (http://grass.osgeo.org/) directly in the processing script (*process*). While in the meteorology field, Williams [WIL 11] used the INTAMAP web service and the 52° North WPS server to interpolate in real time the data reported by sensors via the SOS standard.

In developing countries, where budgets are smaller than those in developed countries, the interest to implement free and interoperable IDSs is all the more important.

Finally, the achieved prototype showed that, technically, it is feasible to implement an intelligent and interoperable GDI with free tools in Morocco. In fact, the use of a service-oriented architecture and FOSS tools, implementing ISO and OGC standards, makes the GDI interoperable with existing systems in public and private institutions. Also, the WPS server integration allows the GDI to not only share reliable data, but also to be interactive, by transforming, with simple clicks, the data into information essential for the decision-making process. It is from all these observations that the interest of this work appears.

3.3.5. *Conclusion*

This research has two objectives: on the one hand, the purpose is to show the importance and the technical feasibility of implementing a free and interoperable GDI in the water and environment sector in Morocco. On the other hand, the major strengths of this system are the online geoprocessing integration and the ease of its use. And this, in order to share, reuses the data and helps in quick decision-making. As a result, these elements will help to better manage water resources, enhance the value of data which represent the Moroccan heritage and optimize the financial expenditures of the public and private sectors.

This work is part of the national e-government strategy, the national water strategy and the 10-95 water law to grant access to public information. It is also aligned with the recommendations of the Economic, Social and Environmental Council.

The aim of this work remains to offer a better decision-making process and, therefore, better integrated and sustainable water and environment management, with all that implies social and economic impacts, affecting in particular the Human Development Index, and including the Kingdom of Morocco in the group of high-growth countries.

3.4. Design of decision support tools

Space-based decision support systems, also known as spatial decision support systems, allow both the dissemination of information and a better knowledge of water resources such as that of a hydrogeological potential map. In addition, such systems incorporate expert knowledge, spatial analysis and/or modeling. Finally, they facilitate the decision-making process. It is in this viewpoint that we thought that our GDI should benefit from the integration of some decision support documents.

3.4.1. *Study area*

The study area is located in southeastern Morocco (Figure 3.18), which is approximately between 5°49 and 3°42 degrees west and between 31°18 and

32°45 degrees north. This area is drained by the upper and middle Ziz and Rheris rivers. The hydrographic basins of these two wadis are limited to the north by the High Moulouya plain and to the south by the Tafilalet plain (Anti-Atlas).

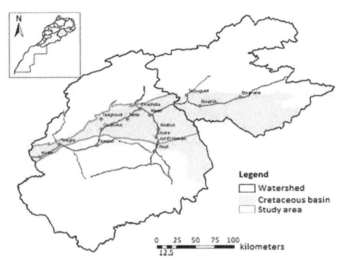

Figure 3.18. *Situation of the study area. For a color version of this figure, see www.iste.co.uk/jarar/spatial.zip*

The Errachidia-Tafilalet region is marked by variable altitudes of 1200 to 800 m which decrease from the north to the south. It is characterized by strong temperature variations and a seasonal distribution of rainfall that is scarce and very irregular.

All the plain is part of the arid Saharan climatic area with cold winters and very hot summers, with a rainfall that barely reaches 60 mm in annual average and low hygrometry. Pluviometry is variable in time and in space. The stations in the study area whose series of measurements is relatively long (1957–2008) represent this variability at the study area level.

Water resources in this region were as follows: 61% of water is perennial, 30% is underground and 9% comes from floods. The inhabitants of the region used water from rivers, floods and springs for their food and agriculture. They have also exploited groundwater (wells) by well-known traditional means in the whole Oughrour region and also by some motor pumps of co-operatives.

3.4.2. Methodology

The applied approach in the framework of this research requires the integration and processing of geographic data relating to several disciplines, such as topography, geology, hydrology and hydrogeology. These information layers were subsequently combined by multicriteria analysis methods to produce a map that represents the hydrogeological potential in the Errachidia region. The use of information layers from the PostgreSQL database will better model the potential areas from a hydrogeological point of view.

The methodology adopted for mapping areas with potential for water resources includes three stages:

– a stage of criteria definition learning;

– criteria classification;

– crossing and evaluation of these criteria.

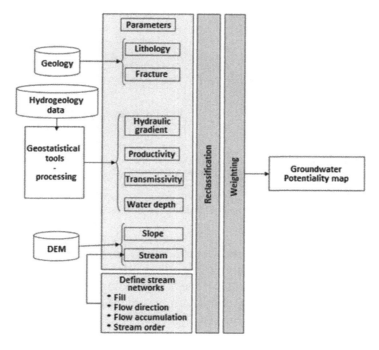

Figure 3.19. *Approach applied for the realization of the hydrogeological potential map. For a color version of this figure, see www.iste.co.uk/jarar/spatial.zip*

3.4.3. *Analysis parameters*

3.4.3.1. *Topography*

The topographic knowledge of a region is an important element in the groundwater resources management. This is a determining factor in the control of run-off rainwater infiltration zones.

In the Errachidia region, topography was represented by a digital elevation model (DEM) (Figure 3.20), generated by digitization and identification of contours and/or spot heights appearing on 20 sheets of topographic maps at a scale of 1/50000.

We chose the TIN (triangulated irregular network) interpolation method as an altitude interpolation algorithm. This makes it possible to produce hydrologically correct DEMs by removing false topographic depressions and generating correct representations of thalwegs and ridges [ELM 03].

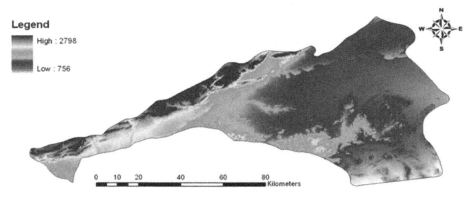

Figure 3.20. *DEM of the study area. For a color version of this figure, see www.iste.co.uk/jarar/spatial.zip*

Subsequently, we applied the slope function on the DEM to arrive at a percentage slope map (Figure 3.21). This plan will subsequently be integrated in the establishment of the hydrogeological potential map (it greatly influences permeability and determines the runoff speed and the flow ratio on infiltration).

A quick review of this information plan shows that the Errachidia cretaceous basin mainly corresponds to a tabular area, with the exception of some hills with slopes ranging between 2 and 10%. In the High-Atlas, the slope is very steep (between 40% and 70%).

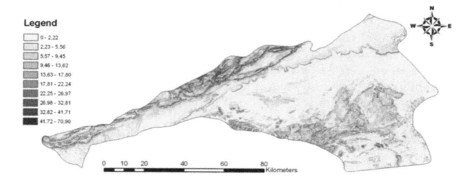

Legend

- 0 - 2.22
- 2.23 - 5.56
- 5.57 - 9.45
- 9.46 - 13.62
- 13.63 - 17.80
- 17.81 - 22.24
- 22.25 - 26.97
- 26.98 - 32.81
- 32.82 - 41.71
- 41.72 - 70.90

0   10   20   40   60   80
                                    Kilometers

Figure 3.21. *Map of the slope of the study area. For a color version of this figure, see www.iste.co.uk/jarar/spatial.zip*

3.4.3.2. *Hydrology*

First, for the automatic extraction of the hydrographic network, we applied on the DEM the flow direction function of the ArcHydroTools application. From the output file, we generated a raster via the ArcHydroTools flow accumulation function. Subsequently, we used an algorithm to automatically extract the virtual hydrographic network of the study area. This network was classified using the Strahler method. This grid will be used as a parameter for the selection of potential areas from a hydrogeological point of view of the infracenomanian aquifer.

3.4.3.3. *Hydrogeology*

In this study, we explored the relational database under PostgreSQL; then we established a geostatistical model of the different hydrogeological parameters of the infracenomanian aquifer.

The geostatisticalal techniques coupled with those of the GIS have provided digital maps of lateral distributions of the hydrogeological variable values. An analysis of these documents highlighted the following points:

– groundwater flow occurs globally for the infracenomanian aquifer from the north to the south (Figure 3.22);

– the depth of the aquifers varies depending on the sectors. The field is shallow, less than 10 m deep, in the western part of the basin and in the Erfoud region. The field tends to deepen in the center and in the eastern part of the basin (50 up to 104 m);

– the average hydraulic gradient (Figure 3.23), comparable to the slope of the piezometric map, decreases overall from upstream to downstream. This slope variation can be explained by known lateral variations of the permeabilities of the reservoirs and by the topographic gradient which determines field gradient.

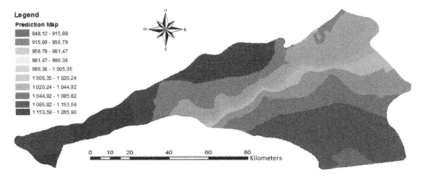

Figure 3.22. *Piezometric map produced by geostatistical modeling. For a color version of this figure, see www.iste.co.uk/jarar/spatial.zip*

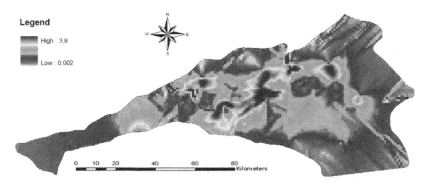

Figure 3.23. *Hydraulic gradient map. For a color version of this figure, see www.iste.co.uk/jarar/spatial.zip*

The productivity corresponds to the flows provided by the works tested by pumping test. The exploitation of the hydrogeological database and the establishment of a thematic analysis on the attribute relating to the flow used during the "Qp" pumping test made it possible to define several productivity classes of the works reported in the Figure 3.24. This map shows that the work flows generally vary between less than 5 l/s and more 100 l/s. The most productive sectors of the field (Q > 40 l/s) are located in the surrounding area and to the west of the city of Errachidia.

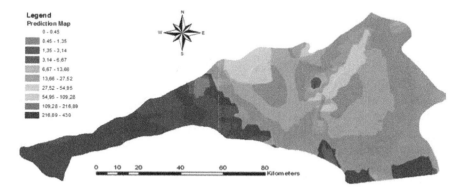

Figure 3.24. *Transmissivity map produced by geostatistical modeling. For a color version of this figure, see www.iste.co.uk/jarar/spatial.zip*

The transmissivity map shows that strong transmissivities align globally according to a NW-SE direction. The most transmissive sectors are located along the Rheris wadi and close to the city of Errachidia. The lowest values are at the level of the banks of the Ziz wadi and the Mayoussel wadi (Figure 3.25).

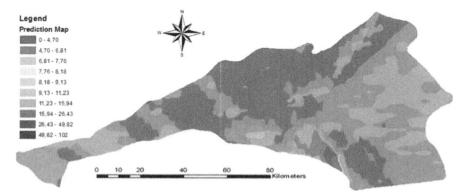

Figure 3.25. *Productivity map produced by geostatistical modeling. For a color version of this figure, see www.iste.co.uk/jarar/spatial.zip*

3.4.4. *Modeling and multicriteria analysis*

In the framework of this study, we used the index modeling method. This method is based on a combination of several information layers (transmissivity, field depth, geology, etc.). A numerical index was assigned to each criterion which gave them a weight; this method is quantitative. Indexed systems methods, with parameters divided into classes, add up the different parameter indexes to arrive at a numerical value reflecting the hydrogeological potential.

The final map (Figure 3.26) is characterized by the presence of seven classes evolving from a "very unfavorable" to a "very favorable" hydrogeological potential degree. Class 1, which characterizes "very unfavorable" areas for the establishment of potential boreholes, appears only rarely on the map. There are therefore six major classes. Classes 2 and 3, which characterize "unfavorable and moderately favorable" areas, are the predominant classes. In fact, the areas of these classes have low transmissivity and productivity. Classes 4 and 5 are characterized by "favorable" areas for the establishment of potential boreholes. This class includes the cities of Errachidia and Goulmima as well as the southern part of the drainage basins of the Ziz and Rheris wadis. In fact, the transmissivity and productivity in these areas are good.

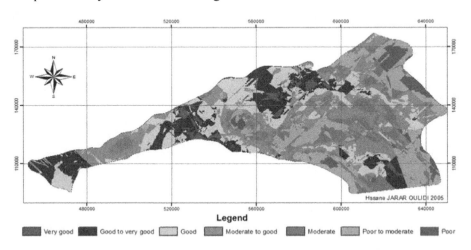

Legend

Very good   Good to very good   Good   Moderate to good   Moderate   Poor to moderate   Poor

Figure 3.26. *Hydrogeological potential map. For a color version of this figure, see www.iste.co.uk/jarar/spatial.zip*

Despite certain benefits of this method, namely facilitating their implementation and cartographic reporting through geographic information systems (GIS), there are difficulties in their development. The main difficulty derives from criteria attribution, their notation and their weight. In fact, each study is different. An important parameter in a study cannot appear in another. In addition, when the number of criteria is high, it is possible that a critical parameter is hidden by the other parameters. Finally, to remedy this problem, we developed a validation approach of this map.

3.4.5. Result validation

The problem of spatial models from GISs is the conformity of the obtained results with ground truth. In fact, to evaluate a thematic map, it is first necessary to have reference data that will be used as an evaluation criterion (Table 3.1 and Figure 3.27). This reference must be terrain data on the works (for example, boreholes flow) to respond to the ground truth concern. From this perspective, when comparing map results and reference boreholes flows, we note that the most productive areas of the map have a very high flow, whereas low productivity areas have a very low flow. Therefore, we can conclude that our approach is valid.

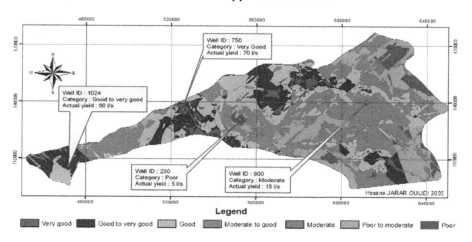

Figure 3.27. *Map that allows validation of the results of the established model. For a color version of this figure, see www.iste.co.uk/jarar/spatial.zip*

Location of reference boreholes		Category obtained from the hydrogeological potential map	Current flow l/s
X	Y		
549170	135225	Poor	3
593740	108770	Poor	5
498440	118556	Poor to moderate	7
516920	123267	Good to very good	60
560766	145008	Very good	70
470176	105148	Moderate to good	20
584682	155879	Very good	75

Table 3.1. *Table comparing the reference boreholes data and the hydrogeological potential map*

3.4.6. *Conclusion and discussion*

For better water resources management, we need detailed, reliable and well-organized information on the state of the hydrogeological environment. With this in mind, the present work made it possible to produce several information plans which will allow better groundwater resources management. GIS and geostatistics were used as tools and techniques to adapt to this problem, while presenting a better multidisciplinary approach for the selection of boreholes establishment sites.

In fact, this technique is not exhaustive and, to enrich it, we propose to carry out a three-dimensional groundwater flow modeling to identify the groundwater path and examine the hydraulic exchanges between the three aquifers of the field.

General Conclusion

This book highlights the importance of the implementation of a water information system in order to better manage and share data and to transmit optimal information to decision-makers and the general public.

In order to respond to this need, it is necessary to establish a relevant technical framework enabling easy and transparent data integration from different data warehouses, providing access to services that can be linked together, in order to process and generate new knowledge and information. The diversity of initiatives at the global scale demonstrates a commitment and a growing need to effectively manage water resources data and ensure the interoperability of shared data.

In this context, this book illustrates to readers the different aspects relating to the implementation of a geospatial data infrastructure conforming with international norms and standards (ISO and OGC). This has been supported by several case studies enabling readers to acquire the experience which constitutes the basis of a water GDI.

Cataloging data on groundwater resources

This case study addressed the hydrogeological data management problem by developing a technical framework for the establishment of a GDI in order to improve access to water resources data in Morocco and to make these data interoperable. Our infrastructure is mainly structured around international standards (OGC and ISO). The data come from different data sources (SaDIN, Badre21, Oracle, etc.). J2E technology has been used for the development of the main components of our application. All document formats used are XML applications, conforming to OGC (GetCapabilities,

GetMap, etc.) as well as W3C standards. The interoperability of our GDI guarantees a high exchange capacity, supporting several dozens of vector file formats and rasters including the main GIS standards of the market. In addition, it also makes it possible to support access to main data, relational and/or spatial databases, and data files of different water resources information providers in Morocco (PostgreSQL/PostGIS, Oracle/Oracle Spatiale, Microsoft SQL Server, Geodatabase, etc.).

This work also led to the development of a cataloging system to make data relating to water resources interoperable and to better manage and preserve them, which allows us to provide reliable information to professionals and decision-makers in the field.

Yet this GDI does not support multiple data models, which may vary from one institution to another. For better interoperability, it is essential to consolidate all existing models in a single map repository. Subsequently, a mediation service will need to be developed which will dynamically convert water resources data from multiple heterogeneous sources to a map repository (common standard). The mediator will manage syntactic, structural and semantic differences between sources so as to isolate client applications from the heterogeneity of the sources. This platform is seen both as a tool for sharing updated data from different data sources and also as a tool for managing water resources on the web and providing the indicators for decision-making.

Sensor Observation Service for sustainable water resources management

Regarding the second case study, we experimented with the possibilities of coupling a GDI with sensors to report information in real time. This system is undeniably a good decision support tool to report piezometric-level measurements in real time. Each measuring station transfers data to the central server, so that later the SOS web service can integrate them at database level. This service is used in this case for real-time monitoring at the piezometric level. The technology can be used to perform various on-site measurements, including groundwater quality.

A hydrogeological data geoprocessing tool integrated into our geospatial data infrastructure

While our geospatial data infrastructure now allows for expanded water resources data access, the exploration and analysis tools (geoprocessing) of these data are still extremely neglected. Against this trend, Khazaz Lamiae's thesis project [KHA 16] aims to produce geoprocessing tools that we would like to share as extensively as possible. This work made it possible to create an interoperable GDI prototype integrating online geoprocessing and interpolation operations with the help of geostatistical methods, using the framework. In the future we intend to combine the developed GDI with scientific models (MODFLOW, HEC-RAS, etc.) for better decision-making as well as add thematic processes in relation to the water field.

Design of decision support tools

The purpose of this case study was to implement a methodology for the development of a decision support document that will be integrated into our GDI. The study uses GIS in combination with geostatistics tools in order to produce a reliable map of potential groundwater resources.

In conclusion, this book represents a modest accumulation of several years of team research in the field and we have been further inspired by the work of our PhD students [MOU 16, KHA 16] and our international colleagues. We hope to have contributed to the enrichment of existing scientific knowledge on the implementation of water GDIs. Thus, we hope that this book will be a practical and relevant pocket tool.

References

[ABA 12] ABADIE A., Formalisation, acquisition et mise en œuvre de connaissances pour l'intégration virtuelle de bases de données géographiques, PhD thesis, University of Paris-Est, France, 2012.

[ALA 12] ALAMI A.E., Système de télémesure des données hydrogéologiques, Report, Département de l'eau, Ministère de l'Energie, des Mines, de l'Eau et de l'Environnement, Morocco, 2012.

[ANC 12] ANCTIL F., ROUSSELLE J., LAUZON N., *Hydrologie : Cheminements de l'eau*, Presses Internationales Polytechnique de Montréal, Canada, 2012.

[ANN 04] ANNONI A., "Infrastructures de données géographiques en Europe", *Geo-Informationssysteme*, no. 4, pp. 241–245, 2004.

[ANN 05] ANNEN A., "Open Geospatial Consortium OGC GML, WMS et WFS", in CAROSIO A. (ed.), *Interopérabilité pour l'utilisation généralisée de la géoinformation*, Ecole polytechnique fédérale de Zurich, Switzerland, 2005.

[ATE 12] ATEMEZING G., TRONCY R., "Vers une meilleure interopérabilité des données géographiques françaises sur le Web de données", *23èmes journées francophones d'ingénierie des connaissances*, Paris, France, 2012.

[AUS 82] AUSTRALIAN GOVERNMENT, Freedom of Information Act, Chief Parliamentary Counsel, Australia, 1982.

[BAL 07] BALLEY S., Aide à la restructuration de données géographiques sur le web : Vers la diffusion à la carte d'information géographique, PhD thesis, University of Paris-Est Marne-la-Vallée, Paris, France, 2007.

[BEA 12] BEAUFILS M., Fusion de données géoréférencées et développement de services interopérables pour l'estimation des besoins en eau à l'échelle des bassins versants, Thesis, Ecole Doctorale Arts et Métiers, CNAM, France, 2012.

[BER 07] BERNIER E., BEDARD Y., Développement de technologies géospatiales. Livrable 3 : Analyse théorique et définition des spécifications d'une infrastructure de découverte et d'accès aux données géospatiales, Report, Université Laval, Canada, 2007.

[BER 11] BERMUDEZ L., ARCTUR D., OGC engineering report: water information services concept development study, OGC, 2011.

[BRÖ 06] BRÖRING A., FÖRSTER T., SIMONIS I., "An integrated software framework for OGC web services", *FOSS4G 2006*, Lausanne, Switzerland, 2006.

[BRÖ 09] BRÖRING A., JÜRRENS E., JIRKA S. *et al.*, "Development of sensor web applications with open source software", *Proceedings of OSGIS: 1st Open Source GIS UK Conference*, University of Nottingham, UK, 22nd June 2009.

[BRO 11] BRODARIC B., SHARPE D., BOISVERT E. *et al.*, "Groundwater information network: recent developments and future directions", *Geohydro2011*, Canada, 2011.

[BRO 16] BRODARIC B., BOOTH N., BOISVERT E. *et al.*, "Groundwater data network interoperability", *Journal of Hydroinformatics*, no. 12, pp. 210–225, 2016.

[BRO 17] BROVELLI M., MINGHINI M., MORENO-SANCHEZ R. *et al.*, "Free and open source software for geospatial applications (FOSS4G) to support Future Earth", *International Journal of Digital Earth*, no. 10, pp. 386–404, 2017.

[BRO 18] BRODARIC B., BOISVERT E., CHERY L. *et al.*, "Enabling global exchange of groundwater data: GroundWaterML2 (GWML2)", *Hydrogeology Journal*, no. 26, pp. 733–741, 2018.

[CAS 13] CASTRONOVA A.M., GOODALL J.L., ELAG M., "Models as web services using the Open Geospatial Consortium (OGC) Web Processing Service (WPS) standard", *Environmental Modelling & Software*, no. 40, pp. 72–83, 2013.

[CUT 98] CUTHBERT A., User Interaction with Geospatial Data, OpenGIS Project, 1998.

[DEL 14] DELIPETREV B., JONOSKI A., SOLOMATINE D., "Development of a web application for water resources based on open source software", *Computers & Geosciences*, no. 43, pp. 35–42, 2014.

[DES 10] DESGAGNE E., Conception et développement d'un SIG 3D dans une approche de service Web : Exemple d'une application en modélisation géologique, Master's thesis, Université Laval, Canada, 2010.

[DON 06] DONNAY J.-P., "Les bases de données spatiales", *Cahiers de la documentation, Bladen voor documentatie*, no. 1, pp. 11–18, 2006.

[DON 09] DONZIER J., WALSHE M., BRÜHL H. *et al.*, Manuel de gestion intégrée des ressources en eau par bassin, Report, Le Ministère français des affaires étrangères et européennes, 2009.

[DUB 07] DUBE E., BADARD T., BEDARD Y., "Service Web de constitution en temps réel de mini-cubes SOLAP pour clients mobiles", *SAGEO'2007*, Saint-Étienne, France, 2007.

[EAU 10] EAUFRANCE, Schéma National des Données sur l'Eau, EauFrance, 2010.

[ELM 03] EL MOJANI Z.E., Conception d'un système d'information à référence spatiale pour la gestion environnementale ; Application à la sélection de sites potentiels de stockage de déchets ménagers et industriels en région semi-aride (Souss, Maroc), PhD thesis, University of Geneva, Switzerland, 2003.

[EUR 07] EUROPEAN PARLIAMENT, "Etablissant une infrastructure d'information géographique dans la Communauté européenne (INSPIRE)", *Journal officiel de l'Union européenne*, no. 108, pp. 1–13, 2007.

[FIT 16] FITCH P., BRODARIC B., STENSON M. *et al.*, "Integrated groundwater data management", in JAKEMAN A., BARRETEAU O., HUNT R. *et al.* (eds), *Integrated Groundwater Management Concepts, Approaches and Challenges*, Springer, Berlin–Heidelberg, 2016.

[FOR 05] FORRER U., "La complexité de l'interopérabilité : compte-rendu de la pratique de la Suisse orientale", in CAROSIO A. (ed.), *Interopérabilité pour l'utilisation généralisée de la géoinformation*, Ecole Polytechnique Fédérale de Zurich, Switzerland, 2005.

[FOR 13] FORTES L.S., Infrastructures de Données Spatiales (IDS) : Manuel pour les Amériques, Report, United Nations, New York, 2013.

[GAR 03] GARBARIN G., *Bases de données*, Eyrolles, Paris, 2003.

[GEO 04] GEOCONNEXIONS, Manuel pour les développeurs de l'ICDG : Produire et publier l'information, les données et les services géographiques, Report, Géoconnexion, Canada, 2004.

[GIU 11a] GIULIANI G., RAY N., LEHMAN A., "Grid-enabled Spatial Data Infrastructure for Environmental Sciences: Challenges and Opportunities", *Future Generation Computer Systems*, no. 27, pp. 292–303, 2011.

[GIU 11b] GIULIANI G., Spatial Data Infrastructures for Environmental Sciences, PhD thesis, University of Geneva, Switzerland, 2011.

[GIU 12] GIULIANI G., NATIVI S., LEHMANN A. *et al.*, "WPS mediation: an approach to process geospatial data on different computing backends", *Computers & Geosciences*, no. 47, pp. 20–33, 2012.

[GIU 14] GIULIANI G., LACROIX P., GUIGOZ Y. *et al.*, Bringing GEOSS services into practice, Workshop, FP7 EnviroGRIDS, University of Geneva, 2014.

[GOG 01] GOGU R.C., CARABIN G., HALLET V. *et al.*, "GIS-based hydrogeological databases and groundwater modelling", *Hydrogeology Journal*, no. 9, pp. 55–569, 2001.

[GUA 03] GUARNIERI F, GARBOLINO E., *Systèmes d'information et risques naturels*, Presses des Mines, Paris, France, 2003.

[HCP 04] HCP, Recensement général de la population et de l'habitat, available at: https://www.hcp.ma/Recensement-General-de-la-Population-et-de-l-Habitat-2004_a92.html, 2004.

[HIN 09] HINGRAY B., PICOUET C., MUSY A., *Hydrologie 2 : Une science pour l'ingénieur*, Presses Polytechniques et Universitaires Romandes, Lausanne, 2009.

[ISO 07] INTERNATIONAL ORGANIZATION FOR STANDARDIZATION, ISO 19136:2007, Geographic information – Geography Markup Language (GML), ISO standard, 2007.

[ISO 11a] INTERNATIONAL ORGANIZATION FOR STANDARDIZATION, ISO 19119, Geographic Information – Services, ISO standard, 2011.

[ISO 11b] INTERNATIONAL ORGANIZATION FOR STANDARDIZATION, ISO 19118:2011, Geographic information – Encoding, ISO standard, 2011.

[ISO 15a] INTERNATIONAL ORGANIZATION FOR STANDARDIZATION, ISO 19103:2015, Geographic information – Conceptual Schema Language, ISO standard, 2015.

[ISO 15b] INTERNATIONAL ORGANIZATION FOR STANDARDIZATION, ISO 19109:2015, Geographic information – Rules for application schema, ISO standard, 2015.

[IST 03] ISTED, Geographic Information Systems and Sustainable Water Management, Ministère de la Transition écologique et solidaire, Paris, France, 2003.

[JAR 09] JARAR OULIDI H., LÖWNER R., BENAABIDATE L. *et al.*, "HydrIS: an open source GIS decision support system for groundwater management (Morocco)", *Journal of Geospatial Information Science*, no. 12, pp. 212–216, 2009.

[JAR 15] JARAR OULIDI H., MOUMEN A., "Towards a Spatial Data Infrastructure and an Integrated Management of Groundwater Resources", *Journal of Geographic Information Systems*, vol. 7, pp 667–676, 2015.

[JON 02] JONES C., PURVES R., RUAS A. *et al.*, "Spatial information retrieval and geographical ontologies an overview of the SPIRIT project", *Proceedings of the 25th Annual International ACM SIGIR Conference on Research and Development in Information Retrieval*, pp. 387–388, Tampere, Finland, 2002.

[JOS 11] JOSÉ M., IGNACIO L., SERGIO V. *et al.*, "Mediterranean water resources in a global change scenario", *Earth-Science Reviews*, no. 105, pp. 121–139, 2011.

[KHA 16] KHAZAZ L., Apport de la géomatique à l'étude de la variation des niveaux pièzométriques : Application à la nappe du Houaz (Maroc), PhD thesis, University of Hassan II Casablanca, Morocco, 2016.

[KOL 03] KOLBOWICZ C., LILLE D., "Informations géographiques interopérables : Application aux îles Loyauté, Nouvelle-Calédonie", *Revue internationale de géomatique*, no. 13, pp. 397–410, 2003.

[KRA 05] KRAAK M.J., SLIWINSKI A., WYTZISK A., "What happens at 52 N? An open source approach to education and research", *22nd International Cartographic Conference*, Spain, 2005.

[KÜN 10] KÜNSTER S., KUHLMANN J., HOLLMANN C. *et al.*, Installation Guide for Sensor Observation Service, Working Group Sensor Web Enablement, 52° North, 2010.

[LAP 12] LAPIEDRA R., DEVECE C., *Introduction to Management Information Systems*, Publicacions de la Universitat Jaume I, Castelló de la Plana, Spain, 2012.

[LON 05] LONGLEY P., *Geographic Information Systems and Science*, John Wiley & Sons, New York, 2005.

[MAK 10] MAKANGA P., JULIAN S., "A review of the status of Spatial Data Infrastructure implementation in Africa", *South African Computer Journal*, no. 24, pp. 18–25, 2010.

[MAS 05] MASSER I., *GIS Worlds: Creating Spatial Data Infrastructures*, ESRI Press, Redlands, 2005.

[MEM 13] MEMEE, Etude prospective de la demande d'énergie à l'horizon 2030, Département de l'Energie et des Mines, Morocco, 2013.

[MIG 12] MIGNARD C., SIGA3D : Modélisation, échange et visualisation d'objets 3D du bâtiment et d'objets urbains géoréférencés ; application aux IFC pour la gestion technique de patrimoine immobilier et urbain, University of Burgundy, France, 2012.

[MIL 13] MILANO M., RUELLAND D., FERNANDEZ S. *et al.*, "Current state of Mediterranean water resources and future trends under climatic and anthropogenic changes", *Hydrological Science Journal*, no. 58, pp. 498–518, 2013.

[MIN 08] MINISTERE DE L'AGRICULTURE ET DES RESSOURCES HYDRAULIQUES, Deuxième projet d'investissement dans le secteur de l'eau - PISEAU II, Tunisia, available at: https://www.afdb.org/fileadmin/uploads/afdb/Documents/Environmental-and-Social-Assessments/30769277-FR-DCPES-PISEAU-II-TUNISIE.PDF, 2008.

[MIN 12] MINO E., BOUAICHA R., JAMIL H. *et al.*, Projet régional de renforcement des systèmes nationaux d'information sur l'eau et d'harmonisation de la collecte des données pour un système partagé d'information sur l'eau : Développement pilote au Maroc état des lieux et plan d'action proposé, Département de l'eau, Unité Technique du SEMIDE, 2012.

[MIS 11] MISSAOUI M., REBAI N., HEIDAR N. *et al.*, "Vers un géo-Framework de catalogage des données aquifères intégrant une infrastructure de données spatiales", *1st International Congress on G.I.S. & Land Management*, Casablanca, Morocco, 2011.

[MOU 16] MOUMEN A., Contribution d'une approche participative et des Infrastructure de Données Spatiales pour la conception d'un système régional d'information sur l'eau : étude de cas au bassin Guir-Ziz-Rhéris et Maider, PhD thesis, Ibn Tofail University, Morocco, 2016.

[MOZ 13] MOZAS M., GHOSN A., État des lieux du secteur de l'eau en Algérie, IPEMED, Paris, 2013.

[NEB 04] NEBERT D., Developing Spatial Data Infrastructures: The SDI Cookbook, Global Spatial Data Infrastructure Association, 2004.

[OLI 05] OLIVEIRA M.A., "Interoperable Geographic Information Services from crisis management perspective", in LI K-J., VANGENOT C. (eds), *Web and Wireless Geographical Information Systems*, Springer, Berlin–Heidelberg, 2005.

[ONE 10] ONEMA (ed.), *Les dispositifs de collecte de données sur l'eau*, EauFrance, 2010.

[ORJ 14] ORJEEBIN-YOUSFAOUI C., *Financer l'accès à l'eau et à l'assainissement en Méditerranée. Les financements innovants : Solution ou illusion ?*, IPMED, Paris, 2014.

[POR 08] PORNON H., YALAMAS P., PELEGRIS E., "Services web géographiques, état de l'art et perspectives", *Géomatique Expert*, no. 65, pp. 44–50, 2008.

[RAJ 01] RAJABIFARD A., WILLIAMSON I.P., "Spatial Data Infrastructures: Concept, SDI Hierarchy and Future Directions", *GEOMATICS'80 Conference*, Tehran, Iran, 2001.

[RAJ 02] RAJABIFARD A., Diffusion of Regional Spatial Data Infrastructures: With Particular Reference to Asia and the Pacific, PhD thesis, University of Melbourne, Australia, 2002.

[REP 11] REPUBLIQUE TUNISIENNE, "Décret-loi no. 2011-41 du 26 mai 2011, relatif à l'accès aux documents administratifs des organismes publics", *Official Journal of the Republic of Tunisia*, available at: http://www.cnudst.rnrt.tn/jortsrc/ 2011/2011f/jo0392011.pdf, 2011.

[ROY 09] ROY G., *Conception de bases de données avec UML*, Presses de l'Université du Québec, Canada, 2009.

[SCH 08] SCHULZ O., JUDEX M., IMPETUS Atlas du Maroc : Résultats de recherche 2000–2007, University of Bonn, Germany, 2008.

[SEM 05] SEMIDE, Technical and financial feasibility studies of the National Water Information Systems in 12 Mediterranean countries, SEMIDE, 2005.

[SLI 05] SLIWINSKI A., SIMONIS I., REMKE A. *et al.*, "Boosting the OGC sensor web enablement initiative by open source web services: the case of 52° North", *AGIT*, Salzburg, Austria, 6–8 July 2005.

[STE 13] STEINIGER S., HUNTER J.S., "The 2012 free and open source GIS software map: a guide to facilitate research, development and adoption", *Environment and Urban Systems*, no. 39, pp. 136–150, 2013.

[TAR 11] TARBOTON D.G., MAIDMENT D., ZASLAVSKY I. *et al.*, "Data interoperability in the hydrologic sciences", *Proceedings of the Environmental Information Management Conference*, University of New Mexico, USA, 2011.

[TAY 12] TAYLOR P., COX S., WALKER G., "Adapting standard information models for water data", *10th International Conference on Hydroinformatics*, Hamburg, Germany, 2012.

[TIR 12] TIRECHE T., KADIRI N., OURAMDHANE A. *et al.*, Vers un système de partage d'informations sur l'environnement (SEIS) – Algérie, Agence Européenne pour l'Environnement, 2012.

[USG 98] US GEOLOGICAL SURVEY, National Water Information System (NWIS), available at: http://pubs.usgs.gov/fs/FS-027-98/, 1998.

[WIL 04] WILLIAMSON I.P., RAJABIFARD A.F. *et al.*, *Developing Spatial Data Infrastructures: from Concept to Reality*, CRC Press, Boca Raton, 2004.

[WIL 11] WILLIAMS M., CORNFORD D., BASTIN L. *et al.*, "Automatic processing, quality assurance and serving of real-time weather data", *Computers & Geosciences*, no. 37, pp. 353–362, 2011.

[WMO 17] WMO, *Good Practice Guidelines for Water Data Management Policy. Australia: World Data Water Initiative*, World Meteorological Organization, 2017.

[WOJ 08] WOJDA P., Hydrogeological data modelling in groundwater studies, PhD thesis, University of Liège, Belgium, 2008.

[WOJ 13] WOJDA P., BROUYÈRE S., "An object-oriented hydrogeological data model for groundwater projects", *Environmental Modelling & Software*, no. 43, pp. 109–123, 2013.

[ZEI 00] ZEILER M., *Modeling our World: the ESRI Guide to Geodatabase Design*, ESRI Press, Redlands, 2000.

Index